Harvesting and Managing Knowledge in Construction

T01 73956

A successful construction business is a knowledge business. And knowledge must be managed effectively to be used efficiently, especially in a complex project-oriented business such as construction, where skills acquired and lessons learned on one project need to be applied to the next.

A holistic approach to knowledge management (KM) is taken in this book to incorporate all of the relevant themes, tackling technological, socio-cultural and organisational issues, with the creation of value as a focus throughout. Information is drawn from a broad range of sources to explain core theories and provide guidance on practical application. Topics covered include:

- changing business relationships in a knowledge economy;
- knowledge creation processes and theories;
- data, text and knowledge-mining techniques;
- the learning construction organisation;
- future technology for knowledge management.

Written by the authors of the first EU-funded KM research project in the field of construction, this textbook is uniquely well-researched, and is the perfect introduction to KM for students across the built environment. It is also a crucial guide to the topic for practitioners.

Yacine Rezgui is an architect by profession, with a PhD in Computer Integrated Construction. He was involved in early developments of product models for construction including STEP while at the Centre Scientifique et Technique du Bâtiment (CSTB) in France. After conducting research on areas relating to knowledge management and collaborative working at Salford University, he joined Cardiff University in 2008, where he is currently affiliated. He is also the Building Research Establishment chair in Sustainable Engineering in areas related to building resilience and adaptability.

John Miles graduated from Manchester University in 1972 as a civil engineer. After some years spent in industry, he obtained his PhD at Birmingham University before being recruited by Cardiff University, where he is currently joint head of the Institute of Machines and Structures. His research there has been in the area of applying artificial intelligence and computational intelligence to the sort of decision-making challenges that routinely arise in engineering and other complex domains.

Harvesting and Managing Knowledge in Construction

From theoretical foundations to business applications

Yacine Rezgui and John Miles

Routledge
Taylor & Francis Group

LONDON AND NEW YORK

Published 2011
by Spon Press

Published 2016 by Routledge
2 Park Square, Milton Park, Abingdon, Oxon, OX14 4RN

Simultaneously published in the USA and Canada
by Routledge
711 Third Avenue, New York, NY 10017

Routledge is an imprint of the Taylor & Francis Group, an informa business

© 2011 Yacine Rezgui and John Miles

The right of Yacine Rezgui and John Miles to be identified as authors of this
work has been asserted by them in accordance with sections 77 and 78 of the
Copyright, Designs and Patents Act 1988.

Typeset in Goudy by Swales & Willis Ltd, Exeter, Devon

All rights reserved. No part of this book may be reprinted or
reproduced or utilised in any form or by any electronic,
mechanical, or other means, now known or hereafter
invented, including photocopying and recording, or in any
information storage or retrieval system, without permission in
writing from the publishers.

The publisher makes no representation, express or implied, with regard
to the accuracy of the information contained in this book and cannot
accept any legal responsibility or liability for any errors or omissions
that may be made.

British Library Cataloguing in Publication Data
A catalogue record for this book is available from the British Library

Library of Congress Cataloging-in-Publication Data
Rezgui, Yacine, 1963–
    Harvesting and managing knowledge in construction : from
    theoretical foundations to business applications / Yacine Rezgui
    and John Miles.
    p. cm.
    1. Knowledge management—Case studies. I. Miles, John, 1950–
    II. Title.
    HD30.2.R494 2011
    624.068'4—dc22
    2010036131

ISBN13: 978–0–415–54595–2 (hbk)
ISBN13: 978–0–415–54596–9 (pbk)
ISBN13: 978–0–203–87609–1 (ebk)

Contents

List of figures   vii
List of tables   ix
Acknowledgments   x
List of abbreviations   xi

1 Introduction   1

2 Changing business relationships   12

3 Construction in the knowledge economy   21

4 Evolution of knowledge management in the construction
   industry   31

5 Knowledge perspectives, approaches and creation processes   41

6 Knowledge management systems   52

7 Domain conceptualisation through ontology   71

8 Construction ontology development   85

9 Complex problem solving: the use of evolutionary algorithms   101

10 Application of genetic algorithms for design   123

11 Future technology for knowledge management   145

12 Knowledge-infused alliances of companies   162

13 Ingredients for a successful knowledge construction organisation   176

14 Value creation: the future of knowledge management in construction 186

References 197
Index 214

Figures

1.1	Authors' information- and KM-related research projects	8
2.1	The trend from traditional, fixed structures to responsive, flexible arrangements	14
3.1	Stakeholder complexity in the construction sector	22
3.2	Key features of the construction industry	23
4.1	Proposed generations of KM in construction	35
6.1	eCognos service architecture	63
6.2	Relationship types in a structured hypertext document	65
8.1	The various stages of the methodology	86
8.2	The eCognos ontology architecture	87
8.3	Method for concept integration making use of a pivotal semantic resource	91
8.4	Snapshot of the eCognos core ontology	95
9.1	An example of a search space	105
9.2	Schematic representation of genetic algorithms	107
9.3a	Design domain	112
9.3b	Problem encoding	112
9.4	Results achieved with two-dimensional genetic operators	116
9.5	Mask-based crossover	119
9.6	Example of change in fitness with generation	121
10.1	An example genotype	126
10.2	Specification of core and atria spaces	131
10.3	Short span structural system background information	133
10.4	Text-based design summary	136
10.5	An example sweep line	138
10.6a	Stage 1 of partitioning	138
10.6b	Stage 2 of partitioning	138
10.7	Adjacency graph	139
10.8a	First partitioning	139
10.8b	Final partitioning	140
10.9	Invalid partitioning	140
10.10	Mutation operator	141
10.11	Crossover operator	141

10.12a	Orthogonal layout	142
10.12b	Orthogonal layout partitioned	142
10.13	Best solution: generation 97	143
11.1	Map of IT research	147
11.2	Provision and maintenance of virtual enterprise projects	156
11.3	Proposed architecture for process-driven integration	157
11.4	The building in its environment	158
11.5	Total lifecycle knowledge perspective on buildings	160
12.1	An alliance in practice	168
12.2	The SME alliance portal architecture	170
13.1	Types of change	178

Tables

4.1	Generations of knowledge management	33
5.1	The comparison between the firm-based and the community-based model of knowledge creation	50
7.1	Key semantic resources in Europe	73
7.2	Product data versus ontology	77
9.1	Areas of interest example in building environment design	102
9.2	Binary and grey encoding	111
9.3	Constraints and criteria	114
9.4	Results for the bending only problem	116
9.5	An example of mutation	120
10.1	Bay dimensions	132
10.2	Evolutionary algorithm tableau for orthogonal building	143
12.1	Issues to be tackled for a holistic understanding of construction alliances	165
12.2	Criteria for sustainable alliances	172
12.3	ICT current situation and envisioned developments	174

Acknowledgments

This book includes findings and results drawn from over 20 research projects completed over the last 20 years, supported by UK (including EPSRC) and European funding councils.

During this period, the authors are fortunate to have collaborated with leading organisations, research institutions and academics/researchers in their field.

This book has been fuelled and inspired by the many discussions the authors have had with their peers in the context of project meetings, conferences or simply informal discussions; many of them will recognise themselves throughout the text.

The book is also shaped by the authors' personal thoughts and reflections that emerged from all these interesting discussions, informed by their own research.

The contribution of everyone is here acknowledged.

Yacine Rezgui and John Miles

Abbreviations

ASP	Application service provider
BIM	Building information modelling
BSI	British Standards Institution
CAD	Computer-aided design
CBR	Case-based reasoning
CSCW	Computer support for co-operative work
DXF	Drawing/data exchange format
EDMS	Electronic document management system
GA	Genetic algorithm
ICT	Information and communication technology
IFC	Industry Foundation Classes
IT	Information technology
KBS	Knowledge-based system
KM	Knowledge management
KMS	Knowledge management systems
NCCTP	Network for Construction Collaboration Technology Providers
OCCS	OmniClass Construction Classification System
OWL	Web Ontology Language
OWL-S	Ontology Web Language for Web Services
RDF	Resource description framework
SME	Small and medium-sized enterprise
SOAP	Simple Object Access Protocol
STEP	Standard for the Exchange of Product Model Data (ISO 10303)
UDDI	Universal Description, Discovery and Integration
VE	Virtual enterprise
WPMS	Web-based project management systems
WSDL	Web Services Description Language
WSMO	Web Service Modelling Ontology

1 Introduction

Why knowledge management? Data, information, and knowledge; Tacit versus explicit knowledge; KM school of thought and practical implications; Positioning of the book; Methodological approach; Structure of the book.

1.1 Why knowledge management?

It is always a challenge to write a new book on Knowledge Management (KM) when the overall feeling is that most has been said and written on the subject. While knowledge management has over the last couple of decades attracted research from management sciences, information systems, and computer science disciplines, it is only recently that companies in the construction industry have started embracing the concept. Internal programmes have been initiated to help promote knowledge sharing practices between staff and across projects. These company-led initiatives have enjoyed different levels of success, and in many cases have experienced failures due mainly to the simplification of the concept of knowledge management, as often reduced to an Information Technology (IT) deployment and diffusion exercise.

The architecture discipline, and more generally the construction industry, has experienced over the last centuries a number of important events: the industrial revolution which has seen the emergence of industrial manufacturing and large scale construction; and, the first and second world wars with their devastating effects on people and the built environment. While the industrial revolution has raised issues related to urban development, industrialisation and society, the first and second world wars raised new challenges related to the need to construct quickly, efficiently, and cost effectively and at the same time develop human friendly architectural interventions.

We are now at the heart of a new revolution, the one of information and knowledge that is reshaping the way we work, live, communicate, and interact within our natural and built environment.

In order to understand the need for knowledge management, it is necessary to see the subject within the broader context of the enormous changes taking place in the global economic framework itself (Neef, 1999). Princeton economist Fritz Machlup highlighted as early as the sixties that there was an increasing proportion of knowledge workers in the workforce (Checkland and Holwell, 1998), coining the phrase 'knowledge industries' in his discussions. In 1993 Peter Drucker, commenting on the manufacturing, service and information sectors wrote:

> We are entering (or have entered) the knowledge society in which the basic economic resource . . . is knowledge . . . and where the knowledge worker will play a central role.
>
> (Drucker, 1993)

Alfred Marshall, a forefather of neo-classical economics, was one of the earliest authors to state explicitly the importance of knowledge in industry and the overall economy:

> Capital consists in a great part of knowledge and organisation . . . knowledge is our most powerful engine of production.
>
> (Marshall, 1965)

The wide introduction of computers in the workplace in the mid-1980s accelerated this paradigm shift. Organisations had the means to quickly capture, codify and disseminate huge amounts of data and information across their subsidiaries and supply chains (Tapscott, 1996). This required adaptations to work practices as these were now better informed by existing information, knowledge and best practice, which is being quickly shared and disseminated. Change and process re-engineering became prerequisites that accompanied and facilitated this paradigm shift.

The widespread deployment of IT in the workplace required new skills and adapted training of the workforce. Organisations began to comprehend the strategic role of information and knowledge and the need for adapted programmes to aspire to the vision of a 'knowledge organisation'. This involved helping employees adapt and respond to change, encourage creativity and innovation, and learn and improve productivity (Neef, 1999).

Knowledge management is now at the heart of modern businesses and economies. It has become an important ingredient to improve business performance and sustain competitiveness. Although Information and Communication Technology (ICT) is central to the implementation of knowledge management initiatives, it is now widely acknowledged that human and organisational factors play an equally important role.

Businesses across all sectors are experiencing an increased and profound shift from tangible to intangible assets centred on people and on the information and knowledge they possess, influenced by developments in information and communications technology. Such changes within the workplace have led to a need to reconsider organisational structures and the role of employees in organisations. Organisations have in their majority initiated in-house programmes to map and re-think their business processes, improve the management of know-how and knowledge, which in turn led to profound workplace and employee management changes reflected in ongoing change management and business re-engineering efforts and initiatives.

More recently, the increasing popularity of knowledge management has been reinforced by the quest for innovation and value creation (Aranda and Molina-Fernandez, 2002; Huseby and Chou, 2003; Vorakulpipat and Rezgui, 2008). KM is perceived as a framework for designing an organisation's goals, structures, and processes so that the organisation can use what it knows to learn and create value for its customers and community (Choo, 2000; Rezgui, 2007b). There is an increasing awareness about the importance of change and change management in sustaining 'leading edge' competitiveness for organisations. The future has only two predictable outcomes – 'change and resistance to change' and the very survival of organisations will depend upon their ability to adapt to and master, these challenges and pave the way to knowledge-infused practices (McAdam and Galloway, 2005; Rezgui, 2007b).

The authors, who have been over the last 20 years actively involved in industry-led research and development initiatives in the construction sector, firmly believe that there is more to report about the subject of knowledge management. In fact, it can be argued that the nature of the construction industry and its unique characteristics deserve a dedicated book on KM that gives a broad but at the same time detailed and industry relevant account on the subject. Also, the literature reveals a lack of directions as to the future of knowledge management in the construction industry. This is an important gap that this book attempts to address.

1.2 Data, information and knowledge

As noted by Alavi and Leidner (2001), it is interesting to reflect on the fact that, in many instances, the epistemological debate about what forms knowledge has been avoided by KM authors by relating and comparing knowledge with information and data. There appears to be no obvious consensus and there have been many definitions on what forms data, information and knowledge.

A commonly held view is that data is raw numbers and facts, information is processed data, and knowledge is authenticated information (Dretske, 1981; Machlup and Mansfield, 1983; Vance, 1997). This view is commonly shared in construction IT circles (Rezgui *et al.*, 2010). However, it can be argued that the presumption of hierarchy from data to information to knowledge with each varying along some dimension such as context, usefulness or interpretability can be misleading (Venters, 2001).

Alavi and Leidner (2001) argue that the effective distinguishing feature between information and knowledge is not found in the content, structure, usefulness or interpretability, but rather: 'knowledge is information possessed in the minds of individuals: it is personalised information (which may or may not be new, unique, useful or accurate) related to facts, procedures, concepts, interpretations, ideas, observations, and judgements'.

Data, information, knowledge and competence may correspond to different levels or forms of human activity (Dahlbom and Mathiassen, 1995). This view considers data as a formalised representation of information, and that information is essentially a charting of knowledge within a shared practice (Venters, 2001). This position can only be sustained by promoting shared practices and experiences of situations.

A significant implication of this view of knowledge is that for individuals to arrive at the same understanding of data or information, they must share a history or context (Alavi and Leidner, 2001). Thus, systems designed to support knowledge in construction organisations may not appear radically different from other forms of information systems used across disciplines, but will be geared toward enabling users to assign meaning to information and to capture some of their knowledge in information and/or data.

An alternative view, would argue that the often assumed hierarchy from data to knowledge is actually inverse (Tuomi, 1999); 'knowledge must exist before information can be formulated and data can be measured to form information' (Alavi and Leidner, 2001). 'Raw data' does not exist *a priori*; thought and knowledge processes are always employed in identifying and collecting even the most elementary data (Venters, 2001).

The authors of this book adopt the view that construction knowledge, drawn from architects and engineers to expert practitioners, has a strong tacit dimension. There is a general agreement that some form of this *knowledge* can be captured and conceptualised into a complex semantic network of *information* (i.e. an ontology), which can then be instantiated on construction projects (i.e. buildings) to form raw *data*. This view is inline with product modelling thinking, such as Industry Foundation Classes (IFC), best reflected in building information modelling (BIM) initiatives.

In fact, it can be argued that knowledge already exists which, when articulated, verbalised and structured, becomes data (Tuomi, 1999). An important aspect to this argument is the fact that knowledge does not exist outside of an agent (a knower); it is shaped by users' needs as well as their knowledge (Alavi and Leidner, 2001; Fahey and Prusak, 1998; Tuomi, 1999).

All the views discussed above will be considered equally in the book so that to encourage open reflections and leave the final interpretations to the reader.

1.3 Tacit versus explicit knowledge

There is an increasing trend to attempt to categorise knowledge with a view to better comprehending, capturing and disseminating all forms of knowledge. Knowledge category models define knowledge into different distinct forms, each of which may be considered and potentially dealt with in specific ways.

Blackler (1995) defines five types of knowledge (encoded, embrained, embodied, encultured and embedded). Through these discrete categories the aim is to reduce the focus on the 'commodification' of knowledge as product, system or service. These categories emphasise the complexity of knowledge within organisations. Alternatively, Boisot (1998) considers knowledge as either codified or uncodified, and as diffused or undiffused within organisations.

Discussion on tacit versus explicit knowledge is probably what has most captivated the KM community. Nonaka and Takeuchi (1995) argue that knowledge creation is not simply a matter of 'processing' objective information. Instead, they believe that it requires the tapping of tacit and often highly subjective insights, intuitions and hunches of individual employees and making such insights available to the organisation as a whole. They argue that making personal knowledge widely available within an organisation is the key to knowledge creation.

In order to achieve this they introduced a model of knowledge creation based on two categories of knowledge which has been previously introduced by Polanyi (Polanyi, 1966): *tacit* and *explicit* knowledge (Routledge, 2000).

Polanyi believed that the 'scientific' account of knowledge as a fully explicit formalisable body of statements did not allow for an adequate account of discovery and growth. In his account of tacit knowledge, knowledge has a subjective dimension: *we know much more than we can tell*. This knowledge is termed 'tacit', while knowledge which we may tell is termed 'explicit' knowledge (Routledge, 2000).

When knowledge is made explicit through language it can be focused for reflection. Polanyi also emphasised the functional aspect of knowledge, i.e. he regards knowledge as a tool by which we either act or gather new knowledge; Sveiby 1997). Sveiby outlines three main theses in Polanyi's concept of knowledge (quoted as in (Sveiby, 1997):

> True discovery cannot be accounted for by a set of articulated rules or algorithms. Knowledge is public and also to a very great extent personal (i.e. it is constructed by humans and therefore contains emotions, 'passion'). The knowledge that underlies the explicit knowledge is more fundamental; all knowledge is either tacit or rooted in tacit knowledge.

In Polanyi's world there is thus no such thing as 'objective knowledge'.

Hence, tacit knowledge is personal, context-specific and therefore hard to formalise and communicate. Explicit or 'codified' knowledge, on the other hand, refers to knowledge that is transmittable in formal, systematic language (Nonaka and Takeuchi, 1995). Tacit and explicit knowledge are not viewed as separate entities, but are rather mutually complementary. Knowledge is created through the social interaction between tacit and explicit knowledge, an interaction that is termed 'knowledge conversion' (Nonaka and Takeuchi, 1995). These issues are further discussed later in the book.

1.4 KM school of thought and practical implications

As discussed earlier, knowledge management is a broad and expanding topic. In reviewing the theory and literature of this field, it is necessary to commit to an identifiable epistemic approach (Rezgui *et al.*, 2010; Venters, 2001). Many such approaches to KM are identified, and have been categorised in various ways (Alavi and Leidner, 2001; Earl, 2001; Schultze, 1998). Schultze (1998) engages Burrell and Morgan's (1979) framework in order to identify a two-fold typology of knowledge within the debate about KM: objectivist and subjectivist. An objectivist approach views knowledge as objects to be discovered. In identifying the existence of knowledge in various forms and locations, technology is employed in the codification of such knowledge objects (Hansen *et al.*, 1999). In contrast, a subjectivist approach suggests knowledge is inherently identified and linked to human experience and the social practice of knowing (Brown and Duguid, 1998). In adopting such a stance, it is contended that knowledge is continuously shaped by the social practice of communities, organisations and institutions.

Conversely, Alavi and Leidner (2001) note that knowledge may be viewed from five different perspectives: (a) state of mind perspective, emphasising knowing and understanding through experience and study (Schubert *et al.*, 1998); (b) object perspective, defining knowledge as a thing to be stored and manipulated (Carlsson *et al.*, 1996; McQueen, 1998; Zack, 1999); (c) process perspective, focusing on knowing and acting (Zack, 1999); (d) condition perspective, emphasising access to knowledge (McQueen, 1998); and (e) capability perspective, viewing knowledge as a capability with the potential for influencing future action (Carlsson *et al.*, 1996).

These different views of knowledge lead to different perspectives of KM (Carlsson *et al.*, 1996):

(a) Information technology perspective, focusing on the use of various technologies to acquire or store knowledge resources (Borghoff and Pareschi, 1998). As knowledge is viewed as an object, KM should focus on ensuring that explicit knowledge, available in the form of knowledge items, is widely accessible across an organisation.
(b) Socialisation perspective, focusing on understanding the organisational nature of KM (Becerra-Fernandez and Sabherwal, 2001; Gold *et al.*, 2001). KM should focus on supporting the processes of sharing, creating and disseminating knowledge.
(c) Information system (IS) perspective, focusing on both IT and organisational capability and emphasising the use of knowledge management systems (Schultze and Leidner, 2002; Tiwana, 2000). The right KM strategy should be put in place to develop and nurture core organisational competencies, and create intellectual capital.

The major implication of these various conceptions of knowledge is that each perspective suggests a different strategy for managing knowledge and a different perspective of the role of systems in support of KM (Carlsson *et al.*, 1996; Alavi and Leidner, 2001). These approaches and strategies will be discussed in later chapters of the book.

1.5 Positioning of the book

Knowledge management has over the years attracted considerable interest from the academic community. A growing number of organisations have included KM into their strategies and have as a result reported (a) business process efficiency improvements, (b) better-organised communities, and (c) higher staff motivation (Nonaka and Takeuchi, 1995; Rezgui *et al.*, 2010).

As discussed earlier, there exist different perspectives and schools of thoughts related to knowledge management. While the preponderance of KM research stems from strategy and organisational theory research, the majority of KM initiatives involve, to a lesser or greater degree, information technology (Alavi and Leidner 2001; Huysman and Wulf, 2006). Such information technology structures are now commonly known as knowledge management systems (KMS). As reported later in the book, existing research on KM in construction tends to focus on the development and adoption of knowledge management systems to different stages of the construction project lifecycle.

Limitations of current approaches to managing knowledge have been constantly reported in the construction and wider literature (Huysman and Wulf, 2006; Rezgui, 2001, 2007b; Sor, 2004), including complex socio-technical, organisational, cultural and political issues.

It is now widely acknowledged that developing the right knowledge management strategies, programmes and systems requires an acute understanding and match between organisational culture (including people, values and information processes) and the external environment (including drivers and barriers) confronting the enterprise.

This book provides broad (technology, people, organisation and society) and unbiased accounts of KM, and explores all different perspectives of knowledge management in the context of the construction industry. It provides a wider and pragmatic perspective on the deployment and adoption of knowledge management programmes and initiatives as well as knowledge management systems, and explores ways in enabling innovation and value creation in the sector.

1.6 Methodological approach

The validity and success of any research initiative critically depends on the appropriate chosen research approach. The latter refers to the principles and procedures of logical thought processes applied to scientific investigations (Fellows and Liu, 2003). The research approach employed, however, will depend to some extent on the research perspective adopted, or the underlying philosophical paradigm. Different researchers will have different thoughts regarding how to conduct their research based on their differing assumptions or views of the world.

In information systems and social sciences research, three broad perspectives have been defined that are the most dominant and provide a theoretical basis for the methodologies employed: positivism, interpretivism (subjectivism), and critical research (Deetz, 1996; Orlikowski and Baroudi, 1991).

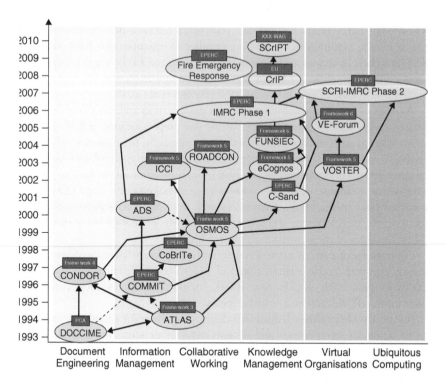

Figure 1.1 Authors' information- and KM-related research projects.

A positivist stance is adopted throughout this book to provide an objective account and reflections on KM research by employing a 'research review' method. This is informed by the authors' work on information and knowledge management, carried out within national and European project consortia, as illustrated in Figure 1.1. Also, it is worth noting that the authors' perspective on KM may differ from that of their peers in the 'construction IT' community in that they would argue that knowledge is primarily and continuously shaped by the social practice of teams, communities and institutions.

This is an interesting debate that we hope will stimulate further discussions and research.

1.7 Structure of the book

The book is structured into 13 chapters addressing important and emerging topics related to knowledge management, ranging from technology to socio-organisational issues. The later chapters of the book have a visionary dimension as they attempt to define the future of the discipline in the construction industry.

Chapter 2 (Changing business relationships) explores the changing nature of businesses and the gradual shift from tangible to intangible assets centred on knowledge and the intellectual capital of an organisation. With the globalisation of the economy, businesses are adopting new ways of delivering products and services facilitated by better-organised supply chains and collaborative networks. A complex business, political and social environment is putting an increasing pressure on organisations to respond to environmental alterations and adapt to them. Organisations need to predict and monitor these transformations and respond with efficient adaptations facilitated by change. Virtual business modes emerge as a result of a desire to improve market position, gain competitive advantages and of course, the will to create value. Organisations worldwide are refocusing strategies to deliver new and existing products and services with maximum value to the customer at the lowest price. In this context, success lies in the ability of an enterprise to combine complementary and distributed (non-collocated) expertise and skills to continuously innovate and remain competitive.

Chapter 3 (Construction in the knowledge economy) refers to construction as a knowledge-intensive industry. This industry involves a large number of very different professions, ranging from design and engineering firms to component/product manufacturers, with ever-growing pressures from clients (building owners) to deliver sophisticated facilities on time and on budget. The chapter provides an overview of the construction industry in terms of structure and characteristics. The barriers to innovation are discussed, as well as knowledge needs of the various stakeholders involved in the design, construction and maintenance stages of a building.

Chapter 4 (Evolution of knowledge management in the construction industry) provides a critical and evolutionary analysis of knowledge management in the construction industry. It spans a large spectrum of KM research published in the management, information systems, and information technology disciplines. An evolutionary KM framework is provided that presents three proposed generations of KM in construction, in terms of three dimensions that factor in (a) the capability of individuals, teams and organisations in the sector, (b) ICT evolution and adoption patterns and (c) construction management philosophies.

Chapter 5 (Knowledge perspectives, approaches and creation processes) gives an account of knowledge management perspectives and approaches. It then focuses on knowledge creation processes and highlights some of the existing conceptual frameworks and their application to the construction sector.

Chapter 6 (Knowledge management systems) discusses a particular type of information systems referred to as knowledge management systems (KMS). The different forms and functionality of KMS are discussed with examples on their application in the construction industry.

Chapter 7 (Domain conceptualisation through ontology) looks at ways in which buildings have been conceptualised and how these conceptualisations have been used to improve data, information and knowledge management capabilities of individuals and teams within projects and organisations. It first discusses the philosophical underpinnings of product data and ontology, and then provides a critical review of their shortcomings, arguing the case for an enhancement of product data to pave the way to more effective, user-friendly conceptualisations of the construction domain through ontology.

Chapter 8 (construction ontology development) describes the role of corporate and project documents in developing a construction domain ontology, taking into account the wealth of existing semantic resources in the construction industry, ranging from dictionaries to thesauri. A modular, architecture-centric, approach is described to structure and develop the ontology. A construction industry standard taxonomy is used to provide the seeds of the ontology, enriched and expanded with additional concepts extracted from large discipline-oriented document bases using information retrieval techniques.

Chapter 9 (Complex problem solving: the use of evolutionary algorithms) covers the basic techniques for solving complex problems using a so-called canonical genetic algorithm. These techniques are interesting in the way they use knowledge to assist engineers and practitioners in their decision-making process.

Chapter 10 (Application of genetic algorithms for design) explores the use and application of genetic algorithm approaches in assisting designers in their architectural interventions. This illustrates the potential of these techniques for the construction industry.

Chapter 11 (Future technology for knowledge management) reflects on several decades of IT development in the construction industry, and attempts to map current efforts with a view of defining directions for the future of knowledge management from a technology perspective. It is argued that semantic (knowledge-infused) process-driven approaches with total lifecycle considerations should underpin ways in which we conceive, design, construct and maintain buildings. The resulting knowledge is shared, reflected upon, improved and re-used with a view to creating value to stakeholders (including clients and occupants).

Chapter 12 (Knowledge-infused alliances of companies) looks into organisational forms to support the vision described in the previous chapter. In particular, the adoption and diffusion of knowledge-driven technologies involve addressing the specificities of small and medium-size enterprises (SMEs) that form a large proportion of the construction workforce. The chapter argues that SME alliance modes of operations promote business process innovation and allow SMEs to compete in new ways, get better reward for their work, gain greater financial strength, which in turn will give them the financial capability to move forward and develop their products and services. A number of concepts are discussed, including: (a) the concept of an SME alliance and its key features, (b) business relationships management in an SME alliance, (c) SME alliance viability and sustainability, (d) the role of information and communication technologies in an alliance, and (e) technical innovation management in an alliance regime of peers.

Chapter 13 (Ingredients for a successful knowledge construction organisation) explores the social, organisational and technology aspects of a construction alliance and highlights important issues that need addressing in order to negotiate the necessary transition from a traditional to a knowledge-driven organisation that engages effectively in knowledge-driven alliances characterised by virtual business modes of operation.

Chapter 14 (Value creation: the future of knowledge management in construction) concludes the book and discusses the future of knowledge management in terms of value creation. Knowledge management processes have inherent value creation capabilities. Five major factors contributing toward value creation are discussed: (a) human networks, (b) social and intellectual capital, (c) corporate social responsibility (d) dustainability, and (e) change processes. Effective knowledge management promotes value creation when it embeds and nurtures the social conditions that bind and bond team members together.

1.8 Targeted audience

The book has a strong theoretical foundation (drawing on state of the art research and theories in the field of knowledge management) as well as a strong practice dimension, informed by numerous case studies and KM deployment initiatives in the architecture, construction and engineering discipline.

The book embodies original research and reflections by the authors aimed at the KM research community. As such the book has a scholar, research, and practice dimension. It is intended for:

- Undergraduate and postgraduate students from the architecture, construction, and engineering disciplines;
- Researchers conducting research in information and knowledge management in these disciplines;
- Practitioners from the same disciplines,

with a view to deepening their understanding of the topic and applying some of the concepts discussed in the book.

2 Changing business relationships

From traditional to virtual means of conducting business; Tangible versus intangible assets; Innovation in knowledge organisations; The importance of 'organisational change'; Conclusion.

2.1 From traditional to virtual means of conducting business

Several factors, including globalisation and the pace of technological innovation, have forced business and industry to adapt to new challenges. These are triggered by an ever more sophisticated society characterised by an increasing demand for customised and high-quality services and products in various segments of industry. Overall, organisations strive to improve their market position, gain competitive advantage and of course, create value out of their tangible and intangible assets. However, as a consequence of this competitive environment, many organisations, including SMEs, operate in a survival mode.

Organisations are now operating in a business environment characterised by (a) a turbulent and instable economic climate, (b) increased competition, (c) consumer and wider societal changes, (d) increased product and service complexity, and (e) organisational drive to sustainable development and corporate social responsibility (Rezgui, 2007b). The prosperity and survival of organisations will depend upon their ability to master these challenges. In this context, organisations require adaptations to factor in a number of business and market requirements, including (Rezgui, 2001):

- *Best of breed:* companies have to provide best of breed solutions which means that they now have to concentrate on their own core competences and outsource for the right quality components/services for the right project.
- *Time to market:* the race in time-to-market necessitates shorter development times which can be facilitated by re-use of out-sourced components.
- *Shorter product cycles:* Increased change and competition necessitate increased agility which can only be achieved by a flexible organisation. Such an organisation concentrates on its core competence and re-invents its offerings.

The past two decades have seen a change in all industries and businesses from organisations that are rigid ('we do it all ourselves') to a more sub-contracting and partnering way of working enabled by better managed co-located, but often distributed global teams. Companies deploy and participate in virtual partnerships in order to effectively sense market demand, develop new products and services, identify new opportunities, improve the quality of their work, and lastly find possibilities to lower organisational costs (Rezgui, *et al.*, 2005).

Amongst the different definitions of the concept of a team (Powell *et al.*, 2004), the following from (Cohen and Bailey, 1997) is one of the most widely accepted:

> A team is a collection of individuals who are independent in their tasks, who share responsibility for outcomes, who see themselves and who are seen by others as an intact social entity embedded in one or more larger social systems, and who manage their relationships across organisational boundaries.

Furthermore, what defines a team is: (a) its unity of purpose, (b) its identity as a social structure and (c) its members' shared responsibility for outcomes (Powell *et al.*, 2004). The distinctive characteristics of virtual teams include the fact that they are geographically, organisationally and/or time dispersed collections of individuals who rely primarily on ICTs to accomplish one or more organisational tasks (Jarvenpaa and Leidner, 1999). They tend to be assembled to respond to specific business and market needs or customer demands.

A number of concepts related to structured forms of virtual teams have been reported in the literature, including: extended enterprise, virtual enterprise, virtual corporation, virtual organisation, supply chain, partnering and alliance (Rezgui and Miles, 2010). While virtual organisation and virtual corporation tend to refer to the same concept, Goranson (1999) draws important differences between the virtual enterprise and virtual organisation (or corporation), noting that the term 'corporation' suggests that there is an inherent vision of corporate identity. Enterprise conveys the meaning that the shared focus is the project at hand. Corporation implies a conventional organisation whose control is centralised. Hence, the virtual enterprise is unified by its mission and distributed goals, not its control system (Goranson, 1999). On the other hand, the extended enterprise extends its activities across a supply chain to offer a complete service or product (Rezgui and Miles, 2010).

In the construction sector, partnering has been introduced as a procurement method to address problems related to conflicts on projects (Black *et al.*, 2000). Partnering implies selection (of partners) on the basis of attitude to team working, ability to innovate and to offer efficient solutions (Egan, 1998). An alliance is a supply-chain driven and strategic form of partnering (Dainty *et al.*, 2001) that sits in between a virtual enterprise and a virtual organisation, as it has a corporate dimension but resembles the virtual enterprise in the way business is conducted.

Collaboration gives rise to the fundamental requirements of labour division into tasks and the coordination of these tasks. The structure of an organisation is reflected in the ways in which it divides its labour into distinct tasks and then achieves coordination among them. Virtual team research to date has focused on the necessity of restructuring traditional organisational structures to exploit the fast development of ICTs (Rezgui, 2007a; Rezgui *et al.*, 2005). In review of the substantial research on team structure in the traditional environment, coordination difficulties facing virtual teams have been found uncounted for (Rezgui and Miles, 2010). The literature relating to the structure of virtual working has put forward some suggestions attempting to achieve high team performance (Kayworth and Leidner, 2000; McDonough *et al.*, 2001). Yet, as managerial structures are associated with poor virtual team performance (Rezgui, 2007a; Vakola and Wilson, 2004), the lack of structures handling virtual team working is an issue for the construction industry, in particular in the context of partnering.

Figure 2.1 illustrates the transformations that businesses are going through. There is a consensus that the traditional form of organisation is too rigid. The trend is to migrate from bureaucratic and hierarchical structures to responsive, flexible arrangements facilitated by ICT. There is a shift from mass production to customisation. The focus is now the service or resulting product as opposed to the organisation. The latter has to be flexible to maximise the quality of the resulting service/product facilitated by collaborative arrangements.

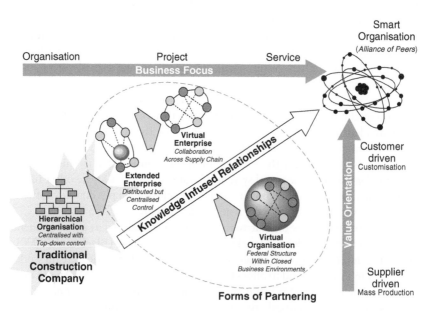

Figure 2.1 The trend from traditional, fixed structures to responsive, flexible arrangements (adapted from Rezgui and Miles, 2010).

2.2 Tangible versus intangible assets

In the traditional organisation there is a tendency to measure organisational wealth and success through annual financial balance sheets and current value in terms of physical assets (buildings, land, equipment, etc.). While these tangible assets are important for the valuation (accountancy perspective), existence and operation of an organisation, there is an increasing awareness that 'employees', the knowledge they possess, and the way in which they use and apply this knowledge (intelligence) is what drives competitiveness and provides the sustainable and innovation dimension of an organisation.

Several authors have tried to define the concept of intangibles. Blair and Wallman (2001) describe intangibles as:

> Non-physical factors that contribute to or are used in producing goods or providing services, or that are expected to future productive benefits for the individuals or firms that control the use of those factors.

A number of authors (Allen, 2003; Blair and Wallman, 2001) have defined intangibles in terms of three factors consisting of human capital, external capital and structure capital. Human capital refers broadly to human competence and expertise. Structural capital refers to the internal processes of an organisation that enables the delivery of its services and products. The external capital relates to the company network of contacts across its value chain, including its customers.

The above factors can be simplified and factored into two main concepts, intellectual capital and social capital.

2.2.1 Intellectual capital

The knowledge a company possesses and the way this is nurtured and applied in business contexts constitute one of its key sustainable advantages. While early knowledge management research has concentrated on knowledge sharing, current research has moved to explore the dynamics of knowledge creation based on corporate knowledge, and how this can be used to meet an organisation's strategic vision and goals (Rezgui *et al.*, 2010). Knowledge management is now perceived by some visionary companies as a mean to define and monitor their strategic objectives and use knowledge to learn and create value for its customers and society (Choo, 2000; Rezgui, 2007b).

Organisations are beginning to comprehend that knowledge and its inter-organisational management, as well as individual and organisational capability building, are becoming crucial factors for gaining and sustaining competitive advantages (Preiss *et al.*, 1996). Moreover, business connections and partnerships to trade knowledge have become of significant importance.

Intellectual capital has enjoyed a very rapid diffusion over recent years and is also a growing area of interest in KM. It encompasses organisational learning, innovation, skills, competencies, expertise and capabilities (Rastogi, 2000). Capabilities are built over a long period and need constant replenishment. Thus, the orientation towards encouraging learning is of prime importance as it requires the continuity of the employment for trained and knowledgeable personnel to safeguard their expertise and best practice and create potential competitive gains and value.

2.2.2 Social capital

Several studies and research have reported knowledge-sharing related success stories where group of employees following their own initiative got together to solve business-related problems. The success of these individually led initiatives leading to some sort of virtual organic structures that did not rely on or even involve senior management staff or directive have gradually attracted interest from both the research community and corporate senior management staff within and outside these organisations. These organic virtual structures, known as 'communities of practice' or 'communities of interest', relate more generally to groups of individuals within or across organisational boundaries that share a common concern, a set of problems, or a passion about a topic, and who deepen their understanding and knowledge of this area by interacting using face-to-face or virtual means (synchronous and asynchronous) on a continuous basis (Wenger *et al.*, 2002).

Social capital emerges from linkages and relationships between individuals and within a community (Choo, 2003). The pattern of linkages and the relationships built through them are the foundation for social capital. Because of its emphasis on collectivism and co-operation rather than individualism, distributed community members will be more inclined to connect and use electronic networks when they are motivated to share knowledge (Huysman and Wulf, 2006). Hence, a focus on social capital in relation to knowledge sharing shifts the attention from individuals sharing knowledge to communities as knowledge-sharing entities (Huysman and Wulf, 2006). In communities, people learn and also invest in the learning of the other members. Therefore, shared understanding and practice is what brings together and nurtures effective communities. However, the key conditions that help communities stay active are mutual trust, respect, a sense of mutuality and recognition by peers (Lesser, 2000); in other words, a high degree of social capital.

Emphasising social capital as the key ingredient to KM also relaxes the managerial and technological bias (Huysman and Wulf, 2006). Technology for KM will likely be more in line with people's opportunity, motivation and ability to share and create knowledge. People will be more inclined to use KM tools when they are motivated, able and have the opportunity to share knowledge with others (Wasko and Faraj, 2005).

Organisational values such as trust, motivation, social cohesion in the workplace are essential to promote knowledge-friendly practices, and infused knowledge in day-to-day practices. In a sense, cultivating the human capital of an organisation contributes to a higher level of social capital. Clearly, it is important to acknowledge social capital when investing in KM and recognise that the higher the level of social capital, the more (distributed) communities are stimulated to connect and share knowledge (Huysman and Wulf, 2006). Moreover, in terms of organisational structure, social capital helps people develop trust, respect, and understanding of others, especially in the context of a rigid and strong organisational bureaucratic culture.

2.3 Innovation in knowledge organisations

It is essential for organisations to develop the right capabilities to leverage their knowledge if they are to compete successfully in future markets. Companies with knowledge management programmes need efficient and effective management of their codified knowledge and best practice (usually maintained within corporate knowledge bases) to ensure creation of new knowledge, its distribution across organisational extent and ready accessibility to all members.

In the general innovation literature, innovation is portrayed as having a number of roles or outcomes: the renewal and enlargement of product/service ranges and their associated markets; new methods of production, supply and distribution; and new organisational and work forms and practices (European Commission, 1996). In the construction literature, Thomas and Bone (2000) identify three key areas for innovation activity that 'can deliver significantly improved quality and value': supply chain management and partnering; value and risk management; and technical innovation.

In the construction sector, innovation is perceived as 'the act of introducing and using new ideas, technologies, products and/or processes aimed at solving problems, viewing things differently, improving efficiency and effectiveness, or enhancing standards of living' (CERF, 2000). It is argued that construction project-based forms of enterprise are inadequately addressed in the innovation literature (Barrett and Sexton, 2006; Gann and Salter, 2000). Project-based organisation focuses on the production and/or delivery side of a firm's business (Artto, 1998), and is characterised by the coexistence of a continuing organisation structure, typically based on functional departments with a temporary organisational structure based on project teams (Barrett and Sexton, 2006; Grant, 1997; Rezgui, 2007a). Organisations across a variety of industries are increasingly experimenting with project-based models of organisation to accommodate and exploit fundamental changes in the nature and roles of markets and technologies (Ayas, 1996; Bonaccorsi *et al.*, 1999; DeFillippi and Arthur, 1998; Kanter, 1997, 1983).

As discussed earlier, a knowledge-based perspective of the organisation has emerged in the strategic management literature (Alavi and Leidner, 2001; Nonaka and Takeuchi, 1995). Organisational knowledge is recognised as a key resource and a variety of perspectives suggest that the ability to marshal and deploy knowledge dispersed across the organisation is an important source of organisational advantage (Teece, 1998; Tsai and Ghoshal, 1998). A key sustainable advantage comes from the way a firm acquires and uses knowledge.

As companies expand there is a limit to the effectiveness of the informal ways knowledge has always been shared within organisations. There is a need for companies to 'know-what-they-know' (Sieloff, 1999). If knowledge is to become a valuable corporate asset it must be accessible, developed and used (Davenport and Prusak, 1998).

The knowledge organisation can be defined through its ability to adapt to the changing environment by creating new knowledge, disseminating it effectively and embodying this knowledge into practice (Nonaka and Takeuchi, 1995). A knowledge-creating company is well placed to provide continuous innovation (Nonaka et al., 2000). Innovation increases when it takes an inter-organisational dimension and knowledge is shared and created across organisations and the supply chain (Ding and Peters, 2000; Levy et al., 2001). In order to achieve such innovation, and when faced with differing knowledge management practices of various organisations, there is a need to establish inter-form collaborative networks which enable discontinuous innovation (Ding and Peters, 2000).

Construction is an interesting example that involves inter-company collaboration. There is a need to introduce forms of knowledge management practice within the construction industry in order improve the partnering process and the way in which knowledge sharing, creation and ultimately innovation are created.

The structure of the organisation has a large impact on its knowledge management success. Early models of knowledge management (Nonaka and Takushi, 1995) build on the interplay between explicit and tacit knowledge at four levels: individual, small group, the organisation and the inter-organisational domain. A departure from this model is required, proposing the n-form organisation focused on the combination of knowledge rather than the division seen in hierarchy (Hedlund, 1994). Such an organisation is characterised by temporary constellations of people, the importance of personnel at local levels, lateral communication, and a catalytic and architectural role for top management, strategies aimed as focusing on economies of depth and hierarchical structures (Hedlund, 1994).

2.4 The importance of 'organisational change'

Organisations find themselves in an almost constant state of change, as they strive to respond to the pressures of the business environment. Drivers for change include:

- Strategic and commercial considerations such as the need to lower costs, improve efficiency, introduce new products and services; such drivers will entail organisational changes including implementing new ways of working, new contractual models, supply chain partnerships, etc.;
- Mergers and acquisitions which involve the bringing together, rationalisation and harmonisation of two or more organisations;
- The availability of new technologies;

- Legislation and changing building regulations;
- Adoption of new forms of partnering.

While the transition from 'traditional' organisations towards more innovative forms of partnering may be accomplished in a limited or partial way through construction projects (involving distributed skills and competencies), when organisational boundaries are being crossed the nature of the changes required internally may be quite large. Issues of confidentiality and security are paramount. It is essential to ensure that knowledge systems are put into place in a way that enables the virtual collaborative work to take place and information and knowledge to be exchanged in a trustworthy fashion. It will be necessary to create management and leadership structures that understand and know how to apply and cope with new forms of organisations, although these may not be in harmony with existing structures within the 'real' organisations. Similarly, communications processes will need to be established, and working styles, roles and rewards may need to be moderated for the new form of organisation to work effectively (Rezgui *et al.*, 2005).

Senior management must therefore adopt a strategic approach to change management, taking into consideration not only at the 'virtual' form of organisation but also at the parts of the 'real' organisation that are likely to be affected and ensuring that the impact of the change is managed in both.

2.5 Conclusion

This chapter presents an overview of the challenges facing construction organisations. Corporate culture and values are in transition, requiring new breed of employees with new skills and capabilities. The composition of today's workforce is changing and has higher expectations than before. Traditional values and behaviours are being challenged. For many years, the prevailing mode of work has been that the manager plans and gives orders to workers who execute the plan and produce. Manageers in the construction industry need to understand that they should give employees at every level a greater voice in organisational operations. An increase in employee participation ultimately fosters stronger employee commitment to the organisation and its business objectives. Whilst we see the emergence of new business relationships and forms of partnering, organisations remain principally human constructs. The management of knowledge and the creation of value involve the intellectual and social capital dimension of an organisation, i.e. their 'intangibles'. Thus the socio-organisational 'equation' consists of a combination of technology, culture and organisation, in which issues including trust, confidentiality, knowledge sharing, etc. must be blended successfully toward the shared organisation purpose. The migration path to successful networking is grounded in human and cultural elements that engage all stakeholders in a manner that is supported by continuous learning. Migration towards any new way of organising and working is an exercise in change, which requires new mechanisms to enable participation and communication.

A number of essential ingredients emerge and should be developed by modern organisations, including the will and ability to:

(a) Increase individual and organisational capabilities.
(b) Develop a culture of individual led initiative and collaboration.
(c) Exploit at best current developments, existing solutions, and advances in Information and Communication Technologies.
(d) Implement change and review and improve business processes and general organisational procedure on a continuous basis.
(e) Implement learning and training strategy aimed at all staff.

In terms of migration to new ways of collaborating and working, a number of issues remain open for the research community. These are discussed in Chapter 13.

3 Construction in the knowledge economy

Structure of the construction industry; The case of small and medium-sized enterprises in construction; Barriers to knowledge sharing and technology adoption; Data, information and knowledge needs and current limitations; The role of research and development; Conclusion.

3.1 Structure of the construction industry

The construction sector is highly fragmented, depending on a large number of very different professions and firms, which are mostly small in size, tend to respond to local market needs and control only one or very few elements of the building design, construction or maintenance process (Rezgui and Miles, 2010). Construction projects involve various professions, including design teams, contractors, facility managers, product manufacturers and suppliers, user associations, clients and investors, and local/regional/national/international (EU) authorities (Figure 3.1).

The increasing complexity of buildings is reflected in the continuous introduction of new procurement paths and methods, construction technologies, materials and construction methods to meet various economic, environmental and societal challenges. This requires the involvement of not only the traditional disciplines (structure, mechanical and electrical, etc.) but also many new professions in areas such as energy, environment, waste and assisted living. For instance, designing a hospital requires not only meeting tighter comfort, energy and carbon requirements but also reducing drastically infection rates by adapted architectural design responses.

The construction industry exhibits characteristics that differentiate it from other industrial sectors (Rezgui and Zarli, 2006), as illustrated in Figure 3.2. These are summarised below:

- It is one of the most geographically dispersed sectors with marked regional differences.
- The industry is project-oriented with a tendency for actors to be involved concurrently in several projects with similar, but sometimes different, roles and responsibilities.

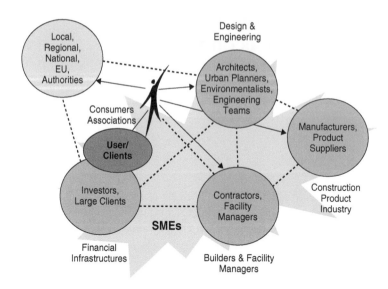

Figure 3.1 Stakeholder complexity in the construction sector.

- The industry is fragmented and structured into a variety of disciplines (e.g., architecture, civil engineering, building services, etc.) that have their dedicated codes of practice, regulations and use their own technical jargon.
- Each construction project is a one-off and unique prototype with distinct characteristics, including choice of construction system, materials, site topography and geology, and local environmental factors. The final product tends to be very durable, lasting several decades and longer, and represents one of the few non-transportable industrial products.
- It is highly regulated. Regulations and standards are more rigorous in construction than in most other industry sectors, with the involvement of several levels of regulatory governance (local, regional, national, European).
- The entry-level for new companies is relatively low because the need for operational capital is small.
- The sector is very labour-intensive, with high mobility of the workforce. Projects now demand an educated and trained workforce as construction technology becomes more sophisticated. The duration of contracts is often linked to the length of the project or design/construction phase.
- Business relationships are temporary and often short-term, bringing together partners who may never work together again.
- Values, norms and cultures tend to vary from one organisation to another, which is reflected in their work practices and business processes. The industry is dominated by SMEs.

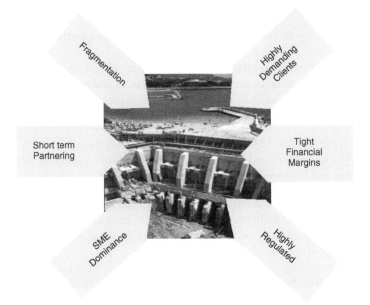

Figure 3.2 Key features of the construction industry.

3.2 The case of small and medium-sized enterprises in construction

The construction industry relies on a very high proportion of small and medium-sized enterprises that operate in every single stage of the design, construction and maintenance process of a building/facility project. Such SMEs tend only to be active in discrete sections of a project and this discontinuity of involvement is a challenge in relation to the adoption of technology and new business models and modes of project operation (Rezgui and Miles, 2010).

The industrialisation of construction poses many unique challenges (Rezgui and Miles, 2010), including:

- Traditional industry process models need continuous adaptation and integration to suit local conditions, new materials and frequently changing business and partners' relationships.
- The expectations and requirements of clients and users vary dramatically from one project to the next. The sector is experiencing the emergence of highly demanding clients and users.
- Impact of new, increasingly complex regulations (environmental, energy, waste, etc.) that must be assessed by specialists to satisfy planning consents and public concerns.
- Concern for quality, timely, to-budget delivery against the threat of financial penalties is causing the major industry players to reduce their circle of specialists and sub-contractors, a majority of which are SMEs.

SMEs face a number of challenges in this environment, including:

- Maintaining and re-using knowledge, adopting best practice and absorbing technology;
- Collaborating across (often) complex value chains for the delivery of increasingly complex and sophisticated services and products;
- The ability to innovate and remain competitive in a fierce business environment that is often controlled by large key player organisations.

Research suggest that SMEs which benefit from interpersonal contacts through networked individuals and teams will gain two kinds of competitive advantages (Bougrain and Haudeville, 2002; Rezgui and Miles, 2010):

- Networks allow SMEs to *decode appropriate flows of information*. They reinforce SMEs' competitiveness by providing them with a window on technological change, sources of technical assistance, market requirements and strategic choices made by other firms.
- Tacit knowledge is very important in innovation. This knowledge cannot be transferred through written documents. It is embodied in *personal knowledge*. Therefore, networks which promote business collaborations become the main transfer channel for knowledge and expertise. Knowing who holds information is crucial.

While SMEs are independent organisations, a large number of them are highly dependent on larger industry players for their work. This can disadvantage SMEs since they can be regarded as only 'piece-workers' to do tasks with little opportunity to add value. They tend to follow conventional business models producing traditional products and services. There is consequently little opportunity for value added services that differentiate SMEs on the grounds of quality of contribution.

The general picture is therefore of an industry that is a pyramid with control being in the hands of large players beneath whom are SMEs who have relatively little influence on the important early decisions that determine project outcomes (Rezgui and Miles, 2010).

The regional character of the construction industry, cultural aspects and the limited vertical and horizontal integration impede the deployment of knowledge-sharing initiatives compared to other industries. This is discussed in the following section.

3.3 Barriers to knowledge sharing and technology adoption

Barriers to knowledge sharing and ICT adoptions in the industry are multifaceted as they relate to organisational, socio-cultural, IPR (intellectual property rights) and technology issues.

From an organisational perspective, there is a lack of long-term partnering between actors that could result in proper knowledge sharing initiatives and cross-projects ICT strategy adoption. In fact, there is no global actor to enforce a structured and effective use of ICT on projects.

From a socio-cultural perspective, there is an overall low awareness about the potential of knowledge-sharing initiatives and the role of ICT (availability and business benefits). There tends to be a low preparedness for change as business relationships are still governed by conventional practices that privilege direct human contacts and reduce the potential for virtual collaboration across the supply chain (Rezgui, 2007a). Most SMEs in the sector do not have a training strategy or programme, and when training is required, it is done in an ad-hoc way.

From an IPR perspective, actors are reluctant to share knowledge and best practice experiences as recognition and reward systems are often not in place, which is exacerbated by the adverserial nature of partnering on projects.

From a purely technical perspective, existing ICT solutions suffer from a number of limitations. For instance, when not developed in-house and acquired off-the-shelf, they tend to have a generic dimension, as they have not been developed for the specific needs of practitioners in the construction sector.

These socio-technical barriers are summarised below:

- Employees do not see any direct value in sharing knowledge and experiences. This is, in fact, perceived as a potential threat to their 'expert' status. The industry needs to address collectively the above barriers to promote a supportive knowledge culture aimed at all staff in the organisation (e.g. by introducing innovative rewards and recognition strategies).
- There is a lack of a clear vision and ICT strategy. The prevailing policy based on acquiring off-the-shelf solutions fails to deliver (Rezgui, 2007b). These commercial solutions tend neither to accommodate existing practices nor build on existing corporate solutions.
- There is a poor software adoption culture as ICT solutions tend to lack scalability as the needs of the organisation and users evolve.
- The sector is fragmented and structured into a variety of disciplines that have their own regulations and use their own technical jargon. There is a lack of shared language holding a common understanding of construction concepts used across disciplines. This prevents effective communication and experience sharing.
- Employees tend to be tied to a physical location (mainly their office) to do their jobs and access software. Access to knowledge from construction sites is essential, but often restricted by network availability or bandwidth limitations.
- Shared knowledge is often not protected in terms of intellectual property rights. Employees feel reluctant to share knowledge when security, confidentiality, and IPR concerns are not properly addressed.

The above barriers are linked to the structure of the industry as noted earlier in the chapter, including:

- The fragmented and discipline-oriented nature of the construction sector;
- The large number of organisations involved in the design and construction of a building facility coupled with the existence of several procurement paths;
- The lifecycle dimension of a construction project with information being produced and updated at different stages in the design and build process with strong information sharing requirement across organisations and life-cycle stages.

3.4 Data, information and knowledge needs and current limitations

Lack of integration across partners is identified as a major issue affecting the performance of the UK construction industry (Egan, 1998; Latham, 1995). ICT has been identified as a key enabler to facilitate sharing of information and knowledge and securing improved performance for the future of the industry. Approaches promoting knowledge management initiatives would help over-come many of the acknowledged barriers to innovation (Rezgui, 2007b; Rezgui and Zarli, 2006).

Knowledge in the construction domain can be classified into the four following categories:

- *Domain knowledge:* this forms the overall industry information context. It includes administrative information, standards and regulations, legislation, codes of practice, technical rules and product/material databases. This knowledge tends to be fragmented, maintained by different institutions (including the Building Research Establishment in the UK) and is made available, in principle, to all companies, through web portals and electronic databases.
- *Organisational knowledge:* this is company specific, and forms part of the corporate intellectual capital. It resides both formally in corporate records and informally through the skilled processes of the organisation. It also comprises knowledge about the personal skills, project experience of the employees and cross-organisational knowledge. The latter covers knowledge involved in business relationships with other partners, including clients, architects, engineering companies, and contractors.
- *Project knowledge:* this forms the potential for usable knowledge and is at the source of much of the knowledge identified above. It is both knowledge each company has about the project and the knowledge that is created by the interaction between organisations (partners on the project). It is not held in a form that promotes re-use (e.g. solutions to technical problems, or in avoiding repeated mistakes), thus companies and partnerships are generally unable to capitalise on this potential for sharing knowledge. It includes both project records and the, recorded and unrecorded, memory of processes, problems and solutions.
- *Individual knowledge:* this is knowledge acquired by individuals through practice drawing from the three above categories of knowledge. This exists

in a tacit form and in several instances is codified but mainly available from users' computers.

Despite the interest and the effort put into knowledge management by many leading companies, the practice of knowledge management in construction is still in its infancy. Many practitioners and researchers have acknowledged the limitations of current approaches to managing the information and knowledge relating to and arising from a project. Among the key reasons for these limitations are (Rezgui, 2001):

- Valuable construction knowledge is acquired over long periods of time through continuous involvement in projects. It resides in the minds of experts working within the domain.
- The intent behind decisions is often not recorded or documented. It requires complex processes to track and record the thousands of ad-hoc messages, phone calls, memos, and conversations that comprise much project-related information.
- People responsible for collecting and archiving project data may not necessarily understand the specific needs of actors who will use it, such as the actors involved in the maintenance of the building(s).
- The data is usually not managed while it is created but instead it is captured and archived at the end of the construction stage. People who have knowledge about the project are likely to have left for another project by this time – their input is not captured.
- Lessons learned are not organised well and are buried in details. It is difficult to compile and disseminate useful knowledge to other projects.
- Many companies maintain historical reports of their projects. Since people always move from one company to another, it is difficult to reach the original report authors who understand the hidden meaning of historical project data. This historical data should include a rich representation of data context, so that it can be used with minimum (or no) consultation.
- ICT provision tends to be fragmented (i.e., application and discipline oriented, dedicated to engineering functions), with little full project lifecycle support.
- Information is conveyed using mainly unstructured documents, which promote the creation and sharing of human-interpretable information. A typical medium-sized project generates thousands of documents.
- New approaches to the management of knowledge within and between firms imply major changes in individual roles and organisational processes. While the potential gains are desired, the necessary changes are resisted.

Addressing the above limitations has the potential to create substantial impacts in the industry by improving learning from experience, reducing mistakes and delivering projects in time and in budget. Also, indirect impact can be created in terms of improved on-site working conditions, health and safety, improved corporate social responsibility and preserving the environment.

3.5 The role of research and development

There is a need to pave the way to a total lifecycle management philosophy in the construction industry (Rezgui and Zarli, 2006). This involves promoting the development of long-term business relationships between stakeholders, while anticipating the expected changes and impact on traditional prevailing management approaches, including procurement methods. This is an area where company-led research and development initiatives can have a major impact as is the case in other industries, including manufacturing.

Research and Development (R&D) efforts and initiatives in construction have traditionally been fragmented and very much subject-orientated, with little emphasis on long-term strategies that would create the right impact to enhance and change practices in industry and lead the way towards sustainable knowledge-driven construction. Moreover, funding bodies in general, and the European Commission in particular, have for over a decade funded project-centred research and technology development efforts in the construction area. While many of these projects have achieved high-quality results, they did not overall create the right impact that would progress the industry from its current state to full adoption of proven and emerging technologies. A few explanations are given below:

- There is a general lack of (construction) industry commitment and involvement in research and development efforts and initiatives. This is often limited to IT managers within R&D departments of large construction organisations with a primary goal of technology watch. This results in a rather poor understanding of the industrial context by the research community, or in some situations, addressing the wrong problems.
- In a general way, consortia set up to carry out funded research did not include the right industry players to produce applicable results (e.g. software vendors).
- Research and development project results tend to be difficult to take up and exploit, as there tend to be a gap with industry average practice.
- There is a lack of inter-programme coordination at a European and national level. Overall, problems tend to be addressed in isolation, with a lack of a strategic and systems-engineering approach.
- There is a lack of inter-project coordination and feedback, co-operation and cross-fertilisation of results across projects and programmes (at a national and European level).
- There is a lack of visionary-led approach within some research programmes. These tend to be fragmented with little emphasis on long-term strategies.
- The human dimension (soft aspects) has been overlooked in a majority of research-funded initiatives that tend to be technology-driven.

Governments are responding to the above issues by launching industry-focused research and development programmes alongside more strategic and long-term actions. This is the case of the Technology Strategy Board in the UK that funds industry-led projects with the potential to involve academic institutions. The latter can secure additional financial support from funding councils such as EPSRC (Engineering and Physical Sciences Research Council). Collaboration between industry and academia is the way forward to accelerate technology adoption in industry. Knowledge management requires an informed intervention as it has a strong socio-technical dimension.

3.6 Conclusion

The building project lifecycle is highly fragmented, structured into discrete stages involving stakeholders with little visibility of the entire project process from concept design to demolition. The main consequence is a lack of information flow across the lifecycle and the supply chain, resulting in inconsistencies with potential major impacts on the final product. This chapter has highlighted the unique features of the construction industry. The dominance of SMEs and the low education profile of the actors in the sector render knowledge management initiatives ever more complex.

It is important that construction organisations address the barriers to knowledge management identified in this chapter. They need to migrate from a centralised approach to knowledge sharing that empowers management staff, to a user- and community-centred approach which involves every single employee of the organisation. The characteristics of the 'participatory' type of culture, with a flat structure, open communication channels, and participation and involvement in decision-making, enhance sharing of information and facilitate communication within and across construction teams. Equally, high motivation levels and job satisfaction are critical success factors.

A construction knowledge-based organisation needs all of its employees to share a culture that promotes the virtues of knowledge capture and sharing. The culture of a company is the set of values, norms and attitudes shared amongst the members of the organisation. A knowledge-based culture requires a number of essential attributes, including:

- A culture of confidence and trust in which people are willing to share the information and knowledge they have, rather than the currently identified 'silos' of isolated knowledge across participating organisations, due in part to the lack of 'systems' for effectively sharing knowledge, coupled with a political climate that engenders mistrust and competition.
- A culture that encourages the learning of lessons from failure as well as success. In many companies failure is associated with blame and can hinder promotion and career progression.
- A culture that recognises that much knowledge is tacit and nurtured in social networks. This recognition places an emphasis on promoting open dialogues between staff so they can develop social links that can promote shared understandings.

- The support for communities of practice where members continuously strive to increase their shared understanding of their collective tasks and to seek continuous improvements in their practice.

Knowledge environment should allow identification, capture and retrieval of relevant knowledge, while promoting and nurturing the social activities that underpin the knowledge-sharing and creation process. As noted in the previous chapter, social capital emerges as an essential condition to enable effective knowledge management. It has been shown that end-users in the construction sector are organised into disciplines that can be assimilated to communities of practice. Members of these communities will be more inclined to use adapted Knowledge Management Systems (KMS) when they are motivated, able and have the opportunity to share knowledge with others. KMS that embed social awareness can play an important role in addressing these requirements, and promote knowledge sharing in the fragmented and distributed networks of the construction industry.

4 Evolution of knowledge management in the construction industry

Generation of knowledge management accounts; Knowledge management generations in construction; Knowledge sharing: first generation of KM in construction; Generation 2: knowledge conceptualisation and nurturing; Generation 3: Knowledge value creation; Conclusion.

4.1 Generation of knowledge management accounts

Several authors have classified knowledge management efforts in terms of discrete generations, and have attempted to envision the future of the discipline. Three accounts of generations of KM emerge (Firestone and McElroy, 2003):

- The first account proposed by Koenig (2002) argues that the first stage of KM evolution focuses on IT-driven KM or knowledge sharing. The use of IT, in particular Internet/Intranet, and tools for knowledge sharing and transfer are perceived as having the potential to create value to the enterprise. Moreover, this stage emphasises the codification of 'best practices' and 'lessons learned'. On the other hand, the second stage focuses on socialisation issues, and factors in human and cultural considerations. This stage stresses the importance of organisation learning applied from the work of Senge (1990), knowledge creation adapted from the SECI model (Nonaka and Takeuchi, 1995), and Communities of Practice (Wenger *et al.*, 2002). This first account suggests that the future generation of KM will focus on taxonomy development and content management.
- The second account is proposed by Snowden (2002) whose first stage emphasises the sharing and transfer of information for decision support. The second stage focuses on processes facilitating tacit/explicit knowledge conversion inspired by the SECI model (Nonaka and Takeuchi, 1995). Snowden (2002) provides a multi-faceted perspective to the future of KM conveyed through the following themes: (a) knowledge is managed as a thing and a view; (b) centralisation of context, narrative and content management; (c) an understanding of organisations as engaged in sensemaking; (d) and scientific management and mechanistic models.

- The third account is proposed by McElroy (1999) who identifies two generations of KM. The first generation focuses on 'supply-side KM' or knowledge
  sharing – It's all about capturing, codifying, and sharing valuable knowledge, and getting the right information to the right people at the right
  time (McElroy, 1999); while his second generation emphasises 'demand-
  side KM' or knowledge creation. While this definition of the evolution of
  KM has received a wider acceptance, Firestone and McElroy (2003) argue
  that this perception of change relates more to the evolution of knowledge
  processing than to knowledge management.

An analysis of the first and second accounts reveals a number of inconsistencies
and does not provide a solid theoretical foundation to the proposed generations
of KM (Firestone and McElroy 2003). The difficulties in Koenig's account begin
in that the first stage makes no reference to technology as a mean to develop
'best practices' and 'lessons learned'. Furthermore, in stage two, his theory does
not provide the connection between (a) CoP and the work of Senge, Non-
aka/Takeuchi, and (b) the connection between CoP and knowledge creation
and innovation. It can be argued that taxonomy development and content
management approaches are already common practice. Moreover, this is part of
the World Wide Web consortium and the Information retrieval sciences community efforts as reflected in the semantic web developments. This, therefore,
represents an extension of the first stage, and should not form the basis of the
envisioned future stage. As to the second account of KM (Snowden, 2002),
the first stage, emphasising information distribution to decision makers, is too
narrow (Firestone and McElroy (2003). It resembles business process re-engineering (BPR), and overlooks the human dimension in knowledge sharing. The
second stage reveals ambiguities between knowledge conversion and knowledge
creation processes. Knowledge conversion in the SECI model represents one
aspect of knowledge creation. In addition, this stage does not discuss the impact
on KM caused by knowledge conversion. The argumentation provided by Firestone and McElroy (2003) raises some serious concerns about Snowden's second
account of KM.

These three generations of KM are summarised in Table 4.1. Despite the difficulties in the first and second accounts, all three accounts provide a level of
similarity: the first generation tends to focus on knowledge sharing, the second generation on knowledge creation. However, the third generation remains
unclear (Firestone and McElroy, 2003).

4.2 Knowledge management generations in construction

It is a challenge to attempt to categorise the evolution of knowledge management in the construction sector as the art of building has evolved for millenaries,
from the start of early civilisations to modern times. Master craftsmen and then
architects/builders perpetuated the tradition of construction from generation
to generation with local adaptations imposed by climate, soil topology, style of

Table 4.1 Generations of knowledge management

	Koenig's account	Snowden's account	McElroy's account
Generation 1	Applying IT to knowledge sharing Best practices and lesson learned	Distributing information to decision support	"Supply-side KM" – knowledge sharing
Generation 2	Human and cultural factors Organisational learning and knowledge creation	Tacit/explicit knowledge conversion	"Demand-side KM" – knowledge creation
Generation 3	Taxonomy development and content management	Knowledge viewed as a thing and a view Centralisation of context, narrative and content management An understanding of organisations as engaged in sense-making Scientific management and mechanistic models	N/A

life, and evolving construction methods. The best examples of codification of knowledge can be found with Vitruvius (*Ten Books on Architecture*) in antiquity and then later with Palladio (*The Four Books of Architecture*) during the Renaissance period. In general, construction knowledge has a very strong tacit dimension nurtured amongst a select number of craftsmen builders with a strong local and regional influence. The advent of computers in the early 1960s gave a ubiquitous dimension to information that enabled its transfer and storage in more than one location.

This chapter provides a categorisation of knowledge management in our information technology era, starting from the first business use of computers in industry. As stated earlier, knowledge may be viewed from five different perspectives (Alavi and Leidner, 2001):

(1) Object perspective, defining knowledge as a thing to be stored and manipulated;
(2) Condition perspective, emphasising access to knowledge;
(3) State of mind perspective, emphasising knowing and understanding through experience and study;
(4) Process perspective, focusing on knowing and acting;
(5) Capability perspective, viewing knowledge as a capability with the potential for influencing future action.

The above perspectives are used to characterise the evolution of knowledge management in construction. Three generations of KM emerge. The first generation adopts an object perspective on KM and emphasises access to data, information and knowledge (Alavi and Leidner 1st and 2nd perspectives). The second generation adopts a process perspective on KM and emphasises knowing and understanding through experience, study and reflection (Alavi and Leidner 3rd and 4th perspectives). The third generation views knowledge as a capability with the potential for influencing future action and creating value for individuals, organisations, and society (Alavi and Leidner 5th perspective).

Furthermore, four distinctive criteria are used to differentiate each generation:

- *Underpinning ICT*: First generation KM is characterised by human interpretable knowledge systems (knowledge is embedded in documents requiring human interpretation). Second generation KM is characterised by semantic-based systems articulated around the use of BIM or ontology. Third generation KM perceives and manages knowledge as an asset (leveraging the intellectual and social capital of an organisation).
- *Socio-technical dimension*: First generation KM is characterised by a strong ICT deployment and adoption dimension, and tends to overlook socio-organisational factors. Human and organisational awareness characterises the second generation KM. The third generation KM is dominated by issues related to social, human and intellectual capital, change management and technology (viewed as an asset).
- *Lifecycle focus*: There is a strong software application focus in the first generation KM. Second generation focuses on making sense of data, information and knowledge in the context of a discipline (e.g. Architecture, structure, and building services). There is a total lifecycle and domain focus in the third generation KM (i.e. integration across architecture, engineering, and construction disciplines and phases).
- *Knowledge perspective*: First generation KM adopts a condition perspective (Alavi and Leidner, 2001), emphasising access to information. Second generation KM adopts a, process perspective focusing on knowing and acting (Alavi and Leidner, 2001). Third generation KM adopts a capability perspective (knowledge promotes capability building with a view to creating value).

Figure 4.1 provides an illustration of the proposed generations of KM in the construction sector using three axes (dimensions). The vertical axis illustrates the evolution of management philosophies in construction. The horizontal axis provides an account of the evolution of ICT solutions. It is worth noting that these technology solutions (a) may in some instances overlap from a chronological perspective, and (b) their sequence of adoption may vary from one organisation to another. The third axis reflects the evolution of the impact of KM on individuals, organisations, and society.

The first KM generation in construction focuses on document-based knowledge sharing initiatives using proprietary or commercial electronic document management systems (EDMS). In this first generation of KM, documents are

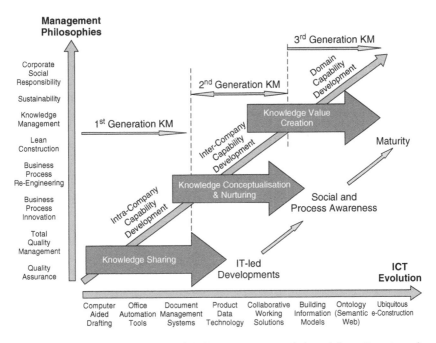

Figure 4.1 Proposed generations of KM in construction (adapted from Rezgui *et al.*, 2010).

perceived as 'black boxes' where the knowledge content requires human interpretation. The production of electronic documents has been facilitated by the widespread and adoption of office automation tools, and the introduction of computer-aided drafting and then computer-aided design in architectural and engineering practices. Documents are archived using techniques drawn from library sciences, mainly based on keywords. This generation of KM can be characterised as technology-driven, and mainly focused on business process automation through IT. Its general aim is to promote intra-company capability development.

The second generation of KM in construction is characterised by efforts aimed at knowledge codification, and conceptualisation of buildings through product-data modelling initiatives, such as STEP (STEP, 1994), the IFCs (IFC, 2010), and more recently, what is known as building information modelling. These efforts have been widely promoted with the advent of the semantic web underpinned by ontologies. This second generation of KM involves smarter means to manage documents, including content indexing, classification (clustering) and retrieval. There is also an increasing social and process awareness about IT adoption and deployment. In a nutshell, there is a shift from intra- to inter-company capability development, with a focus on projects.

The third generation KM is initiated with the emergence of environmental and societal concerns and the drive to deliver human and environmental friendly buildings and a sustainable built environment. This is where organisations start considering their corporate social responsibility. There is also a greater awareness about the human and soft dimension in relation to employees, knowledge and processes, and a shift from tangible to intangible assets in the way the organisation is perceived and managed. This is a generation of knowledge maturity with a view of creating value out of knowledge for the benefit of individuals, organisations, and society.

These three generations of KM in construction are described in further details in the following sections, drawing on existing KM core literature.

4.3 Knowledge sharing: first generation of KM in construction

The evolution of knowledge management in construction has to some extent reflected the evolution of IT adoption in the sector. Prior to the introduction of IT, and up to the early 1980s, the main concern of the industry was project information management (Rezgui *et al.*, 2010). As a larger construction project may eventually result in the production of tens of thousands of documents, the concern of the industry was to provide easy and quick means to locate the appropriate document(s) (Löwnertz, 1998). This relied mainly on managing project documentation either on an ad-hoc basis or at best using traditional library archival methods.

Documents are perceived as information and knowledge carriers (initially available in paper form); they can simultaneously be text-based and/or provide graphical information for a particular purpose. Central to the idea of a document is usually that it can be easily transferred, stored and handled as a unit (Löwnertz, 1998; Rezgui *et al.*, 1998). In fact, project documents have remained similar for the last decades. Drawings (plans, sections, elevations, etc.) and text-based documents (bills of quantities, specifications, etc.) look much the same in terms of contents and form. However, the process involved in producing, distributing/sharing, approving, and updating these documents has evolved as a result of the introduction of IT (Löwnertz, 1998). A mixture of different document management methods is today used across projects and organisations (Hjelt and Bjork, 2006; Boddy *et al.*, 2007).

Initial document management systems used basic file management capabilities found in operating systems, and included a number of functionality (Hjelt and Bjork, 2006) such as: user authentication; a main retrieval mechanism based on either hierarchical folders or meta-data; handling of revisions and change management; viewing of CAD-files using special purpose software; full text search capability. In a nutshell, documents are stored centrally on a server and users interact with this central repository through simple interfaces and then later, with the advent of the Internet, through simple web browsers. Some of these systems have been developed in-house or offered by third parties as ASP-services (application service provider). EDMS tend to treat documents as black boxes, while capturing meta-information, which enables humans or document management systems to search for, retrieve and open documents. Meta-data was initially included in documents or cover pages, drawing headers, etc., and then directly stored in the database that underpins the EDMS.

In parallel with, and related to, document management efforts, it was the increasing use of CAD facilities in design offices from the early 1980s which prompted the first efforts in electronic integration and sharing of building information and data (Bjork, 1989; Eastman and Siabiris, 1995). Here, the ability to share design data and drawings electronically either through proprietary drawing formats or via later de facto standards such as DXF (Drawing/Data Exchange Format), together with the added dimension of drawing layering had substantial impacts on business processes and workflows in the construction industry (Boddy *et al.*, 2007). Although in these early efforts, sharing and integration was mainly limited to geometrical information, effectively the use of CAD files was evolving towards communicating information about a building in ways that a manually draughted or plotted drawing could not (Bjork, 1989; Eastman and Siabiris, 1995).

Also, related to the above, several library management-led efforts have been carried out with a view to standardising and developing a shared terminology and language (Ekholm, 1996). This involved the development of a variety of semantic resources ranging from domain dictionaries and classification systems to specialised taxonomies (Rezgui, 2007c). The most notable efforts include the BS 6100 Glossary of Building and Civil Engineering terms produced by the British Standards Institution (BSI, the independent national body responsible for preparing British Standards) and the OmniClass Construction Classification System (OCCS – http://www.occsnet.org/), developed in Canada by the Construction Specifications Institute. The latter addresses the construction industry's information management needs through a coordinated classification system.

Despite the tendency to emphasise the role of IT in KM, there is during this first generation of KM an increase of powerful arguments for a more holistic view which recognises the interplay between social and technical factors (Pan and Scarbrough, 1998). For instance motivation and trust have been identified as important ingredients to promote a knowledge-sharing culture (Rezgui, 2007a). In fact, the raising of awareness about the influence of people and processes, combined with the quest for enabling computers to make sense of data and knowledge and support more effectively business process automation, are what signal the start of a new, second generation of KM in construction.

4.4 Generation 2: knowledge conceptualisation and nurturing

The introduction of object-oriented CAD in the early 1990s by companies such as AutoDesk, GraphiSoft and Bentley Systems has provided new insights into knowledge management. Data 'objects' in these systems (doors, walls, windows, roofs, plant and equipment, etc.) stored non-graphical data about a building and the third party components which it comprises as 'product data', in a logical structure together with the graphical representation of the building (Anumba and Evbuomwan, 1999; Wetherill *et al.*, 2007). These systems often supported geometrical modelling of the building in three dimensions, which helped automate many of the draughting tasks required to produce engineering drawings.

When combined with the increasing ubiquity of electronic networking and the Internet, this allowed many companies to collaborate and share building information and data which in turn led to new ways of communicating and working (Bjork, 1989).

It became clear that in order to take best advantage of the potential for CAD and object/product model integration, there was a need for more coordinated standards which would simplify and encourage its uptake (Eastman and Siabiris, 1995). These standards defining efforts came in the form of the STEP application protocols for construction (STEP, 1994). This work, inspired by previous work primarily in aerospace and automotive fields, formed part of ISO 10303, the International Standard for the Exchange of Product Model Data. Latterly, the International Alliance for Interoperability defined the Industry Foundation Classes (IFC, 2010), a set of model constructs for the description of building elements.

Preceding and in some cases concurrent with this work, the research community produced several integrated model definitions including ATLAS (Bohms *et al.*, 1994), the RATAS model (Bjork, 1994) and the COMBINE integrated data model 1 and 2 (Augenbroe, 1994, 1995). These research efforts tended to propose a data model and also provide a suite of tools to manipulate the model (as a proof of concept), or a central database to serve model elements to other applications used in the project process via some form of adapter. Several incarnations of the central database idea have been produced using IFC Model Servers (Boddy *et al.*, 2007) designed to host entire building models described in the IFC format.

Within the last three to four years, researchers and commercial application developers in the construction domain have started to develop tools to manipulate complex building models. By storing and managing building information as databases, BIM solutions can capture, manage and present data in ways that are appropriate for the user of that data. Because the information is stored in a logically centralised database, any changes in building information data can be logically propagated and managed by software throughout the project lifecycle (Howard and Bjork, 2008). Building information modelling solutions add the management of relationships between building components beyond the object-level information in object-oriented CAD solutions. This allows information about design intent to be captured in the design process. The building information model contains not only a list of building components and locations but also the relationships that are intended between those objects (Boddy *et al.*, 2007).

While a great deal of effort has been invested in the conceptualisation and formal description of buildings, the capture of tacit (or conversion from tacit to explicit) knowledge has remained problematic. Tacit knowledge is in fact implicit and difficult to share as it is acquired through experience, learnt by doing and apprenticeship (Nonaka *et al.*, 2000). Practitioners have started realising that to succeed in sharing tacit knowledge, it is necessary to share knowledge through know-how involving face-to-face or virtual interaction between knowledge transmitters and receivers. This is where the concept of communities of practice (Wenger *et al.*, 2002) started gaining popularity in construction circles, reflected by the emergence of discussion forums initiated by users within and across organisations.

This second generation KM has also seen the emergence of ontologies. Various definitions of what forms an ontology have been formulated and have evolved over time. A good description of these can be found in (Corcho *et al.*, 2003). From the authors' perspective, the best definition that captures the essence of an ontology is the one given by Gruber (Gruber, 1995): 'an ontology is a formal, explicit specification of a shared conceptualisation'. As elaborated in (Studer *et al.*, 1998): 'Conceptualisation refers to an abstract model of some phenomenon in the world which identifies the relevant concepts of that phenomenon. Explicit means that the types of concepts used and the constraints on their use are explicitly defined. Formal refers to the fact that the ontology should be machine processable'.

The second stage of KM ends with an increased awareness about the 'soft' dimension of KM, and the importance of the human and organisational factors. This paves the way to a new generation of KM: knowledge value creation.

4.5 Generation 3: knowledge value creation

While knowledge sharing and nurturing are perceived as first and second (present) generations of KM, the authors argue that for the future of KM, it is necessary to explore and emphasise the impact of KM related initiatives on people, organisations and society in terms of value creation (Rezgui, 2007a).

The relationship between value creation and KM has been discussed extensively in recent literature (Chase, 1997; Liebowitz and Suen, 2000; Rezgui, 2007b).

There are signs that suggest that the construction industry has started comprehending the concept of value creation (Rezgui, 2007b), and has thus entered the third generation of knowledge management. This is best illustrated by efforts to (a) promote the sustainability of our built environment, (b) promote health and safety on projects, and (c) increase the awareness of construction companies on their corporate social responsibilities.

In relation to point (a), while there are many new techniques, tools and practices being used and experimented to promote sustainability, the collective learning and advancement of sustainability issues is hampered by an inability to identify, capture, manage and re-use the large amount of potential knowledge generated by individuals across organisations and projects.

Point (b) recalls the fact that the construction industry carries a bad reputation in terms of health and safety incidents. It is important to better identify, assess and predict hazards and deal with them at an early stage. This, for instance, improves designers' awareness and understanding of the safety implications of their decisions.

Point (c) suggests that construction companies are now assessing their carbon footprint (linked to their activities) and have started rethinking and redefining their image, driven by recent work on corporate social responsibility (Rezgui, 2007b).

All the above issues are robust indicators that the industry has now entered a new form of management of knowledge centred on value creation and management.

The final chapter of the book discusses and elaborates further on the proposed third generation of KM.

4.6 Conclusion

The chapter discusses three generations of KM in the construction industry. It is argued that the third generation of KM focuses on and revolves around value creation. In fact, once knowledge is shared and created, it is necessary to establish the 'impact' this has on people, communities and society. Although the current design of organisational KM processes tends to be fairly prescriptive, with an approach governed by quality, time and cost, but with little emphasis on innovation and value creation, organisations have the potential of introducing a paradigm shift, promoting customer centred and product/service performance-driven development approaches where intangibles such as intellectual capital, customer and society values transcend the traditional practices that concur to the delivery of product and services.

Performance-driven processes place the client in the centre of organisational dynamics with a commitment to ensure full satisfaction and value to customers, employees and, ultimately, the society.

5 Knowledge perspectives, approaches and creation processes

Perspectives to knowledge management; Knowledge management approaches; Knowledge creation processes; Conclusion.

5.1 Perspectives on knowledge management

Several authors have attempted to provide an account of knowledge management perspectives or school of thoughts. These include Earl's schools for knowledge management (Earl, 2001), McAdam and McCreedy's categories of model for Knowledge Management (McAdam and McCreedy, 1999), Schultze's perspectives on knowledge management (Schultze, 1998), Alavi and Leidner's Perspectives on Knowledge Management (Alavi and Leidner, 2001). These KM perspectives are interesting in that they aim at providing a broad understanding of the discipline and its applications in business and industry. A brief overview of these perspectives is given below.

5.1.1 Earl's schools of knowledge management

Earl (2001) maps the different types of knowledge management strategy developed by organisations. Seven schools of knowledge management strategy are identified: systems, cartographic, engineering, commercial, organisational, spatial and strategic (Earl, 2001). These schools of KM are organised into three categories: technocratic, economic and behavioural.

The technocratic school perceives knowledge management as employing information technology to assist employees in their work and make knowledge available to them through different means. It is described in terms of three dimensions: systems, cartographic and engineering. The systems dimension treats knowledge as explicit objects which can be captured and stored. The cartographic dimension focuses on mapping organisational knowledge and allows employees to locate and access the right expertise across the organisation. This is adapted to non-codified tacit knowledge. The engineering dimension focuses on the process of using knowledge to complete tasks or activities through a process of learning and doing.

The economic school of thought, also known as the commercial school, focuses on using the organisation's knowledge to produce revenue streams.

The exploitation of knowledge as an asset is perceived as a revenue-creating activity.

The behavioural school approaches knowledge management from a behavioural perspective, stimulating and orchestrating managers and managements to proactively create, share and use knowledge resources. It is described in terms of three dimensions: organisational, spatial and strategic. The organisational dimension focuses on 'organisational structures or networks sharing or pooling knowledge'. The spatial dimension views individuals in the centre of knowledge activities and promotes face-to-face rather than virtual (technology-oriented) means of communication. The strategic school perceives value as knowledge-based products and services. It draws on aspects of the other schools to create the value dimension of knowledge.

Earl's model tends to be criticised for its simplification of the social dimension of knowledge sharing and creation, as this is only considered in the context of the spatial school. However, it does include a strategic dimension of KM and suggests a value dimension of KM. This is in line with the third generation of KM in construction presented in the previous chapter.

5.1.2 McAdam and McCreedy's categories of model for knowledge management

McAdam and McCreedy (1999) provide an alternative structure for understanding knowledge management. Three categories of model for knowledge management are proposed; intellectual capital models, knowledge category models and social constructionist models.

The intellectual capital model views knowledge in a scientific way, much like the technocratic school described in Earl's schools of knowledge management. This perspective treats knowledge like an asset and focuses more on explicit knowledge which can be easily stored and shared.

The knowledge category model describes the process of converting knowledge from its tacit form to an explicit form or vice versa. This model uses a combination of social and mechanistic methods of transferring and sharing knowledge.

The social constructionist model focuses more on the social aspects of knowledge and the learning process within an organisation. It is concerned with how knowledge is socially constructed within an organisation, and views knowledge management as a social process.

5.1.3 Schultze's perspectives on knowledge management

Schultze's perspective (Schultze, 1998) is adapted from a framework developed by Burrell and Morgan (Burrell and Morgan, 1979). It identifies two paradigms, functionalist and interpretive (Schultze, 1998):

- *Functionalist paradigm*: The functionalist paradigm views knowledge as objects that can have an explicit or tacit dimension. These objects can be 'captured, manipulated, transferred, and protected' (Schultze, 1998). This is in line with (Hedlund, 1994) and (Nonaka and Takeuchi, 1995) theory of knowledge. Moreover, this perspective provides an objective representation of the world in which knowledge exists in a variety of forms and locations that can be manipulated by human agents.

- *Interpretive paradigm*: This perspective views knowledge management as a process that is strongly related to human experience and social practices of knowing (Schultze, 1998). In this perspective, knowledge is nurtured by the social practice of communities.

5.1.4 *Alavi and Leidner's perspectives on knowledge management*

Alavi and Leidner (2001) note that knowledge may be viewed from five different perspectives: state of mind, object, process, access to information, and capability.

The state of mind perspective focuses on people using information to enhance their own knowledge and understanding through experience and study. This process is concerned with 'access to sources of knowledge rather than knowledge itself'.

The object perspective considers knowledge as an object which can be easily stored and manipulated. This perspective deals primarily with explicit knowledge and examples of stored knowledge include databases, organisational intranets and CD-ROMs.

The process perspective focuses on 'knowledge being a process of applying expertise'. This perspective is concerned with 'the process of creating, sharing, and distributing knowledge'.

The access to information perspective is related to the object perspective. They both view knowledge as an object, but whereas the object perspective is focused on storing and manipulating knowledge, the access to information perspective is focused on making the knowledge easily accessible. Examples of this perspective in a practical environment would be a search engine which could be used to search and retrieve information. The capability perspective describes knowledge management as 'building core competencies and understanding strategic know-how'.

5.2 Knowledge management approaches

There exist several approaches to knowledge management that involve a wide range of social and process-oriented techniques. A number of these are briefly described below.

5.2.1 *Sensemaking*

Sensemaking involves making sense of the unknown by introducing structure and order, and exploring how individuals construct realities, and the effect this has on their environmentt (Weick, 1995, 1996). Weick (1995) outlines sensemaking as a process that is:

- *Grounded in identity construction*: The sensemaking process involves the construction of self's identity.
- *Retrospective*: Sensemaking involves past experiences and events. The sensemaker's reality is interpreted and historic.
- *Enactive of sensible environments*: the sensemaker influences and at the same time is constrained by the environment in which he or she operates.
- *Social*: the sensemaking process has a strong social dimension: 'Conduct is contingent upon the conduct of others, whether those others are imagined or physically present'.
- *Ongoing*: Sensemaking is a continuous process constantly shaped by individuals interpreting and making sense of situations.
- *Focused on and by extracted cues*: Individuals are in continuous search of objective or subjective cues to elicit information about, make sense of and update situations.
- *Driven by plausibility rather than accuracy*: In order to make sense of a situation an individual needs sufficient and credible information to carry out the task in hand.

The relationship between sensemaking and knowledge management is obvious from the above attributes. In fact, the organisation, and the individuals that compose it, is perceived as a system of perception, engaged in continuous sensemaking. Individuals create and share knowledge within their environment and by doing so construct and express their own identity, and as a result, have an effect on their environment. Knowledge management is thus perceived as a social process continuously influenced by individuals, their knowledge and the ways this is shared and used in their environment, i.e. team or organisation. Making sense of situations influences action and triggers a decision-making process. It is thus important to understand the relationship between sensemaking and action.

5.2.2 Requisite variety

The law of requisite variety states that only variety can destroy variety (Ashby, 1956). In fact, while carrying out their tasks individuals must effectively manage the complexity they face (Espejo, 1993). This complexity involves the internal and external environment of the organisation. It is important for the internal diversity of an organisation to match the variety of the external diversity in order to cope with changes in this environment (Ashby, 1956). More importantly, organisations need to find a balance between factors that attenuate environmental complexity and those that amplify managerial complexity. However, modern organisations involve too much amplification, and too little attenuation (Espejo, 1993).

A parallel can easily be drawn with knowledge management systems. These tend to focus on informational needs of individuals, while the environment in which such information is provided is often overlooked. A greater emphasis should be given to the processes that underpin knowledge activities, as opposed to the knowledge itself.

Change resulting from the adoption of knowledge management systems is thus determined by the internal coherence of the system and not by information about external events (Espejo, 1993). To some extent, change may be triggered by knowledge, but is not determined by it. Equally, it is not the content of the information that determines the individual's response, but rather the organisational structure into which knowledge is nurtured.

The development and deployment of knowledge management programmes should involve a balance between observations of the external informational environment, the structure into which such observations are attenuated and the amplification of individuals' action in order to initiate change (Nonaka and Takeuchi, 1995). Approaches that are limited to the codification of individuals' knowledge are insufficient. Knowledge is dynamic and evolving, reflected by the complex changing environment in which it is created and nurtured. It is thus important to promote a culture of challenging assumptions, and supporting individuals in reflecting upon their knowledge and its relevance to new situations (Venters, 2001).

5.2.3 *Reflection in action*

While there is a general belief that knowledge exists and is simply applied to a situation, for instance for a problem-serving purpose, Schön (1983) proposes a notion of knowledge-in-action by which the process of acting reveals a 'kind of knowing' that does not involve prior knowledge or intellectual operation. In arguing this view, Schön suggests that 'knowing' has the following attributes:

- There are actions, recognitions, and judgements which we know how to carry out spontaneously; we do not have to think about them prior to or during their performance.
- We are often unaware of having learned to do these things, we simply find ourselves doing them.
- We may or may not be aware of internalised understandings during action which reveal difficult to describe.

Knowledge in action is hence applied when the situation faced is familiar. However, when individuals are faced with a new situation, a 'breakdown' occurs (Winograd and Flores, 1986). Individuals have to reflect on what they are doing and must analyse the new situation and the relevance to it of their knowing. In other words, they have to think on how to act upon the new situation. This process is defined as reflection-in-action. Through reflection the individual is able to 'surface and criticise the tacit understandings that have grown up around the repetitive experiences of a specialised practice, and can make new sense of the situations of uncertainty or uniqueness which he may allow himself to experience' (Schön, 1982). When facing new situations individuals may feel uncomfortable as they cannot justify the quality or rigour of their actions (Schön, 1982). In this context, individuals need to analyse and make sense of the situation and consider alternative approaches to address the problem.

In summary, there are two sorts of reflection: reflection-in-action and reflection-on-action. The former involves a level of awareness throughout the action, while in the latter reflection is done after the action or event. The value of a person's reflection can be greatly enhanced by a greater understanding of the process. This allows the participants to create their own knowledge and theory relevant to their own specific situation.

5.2.4 Problem-structuring methods

Problem-structuring methods are a set of approaches that enable problem owners to manage their problem situations (Rosenhead and Mingers, 2001). These approaches are underpinned by a variety of theoretical frameworks and are briefly described below:

- *Strategic Options Development and Analysis* (SODA) (Eden and Ackermann, 1998) is based upon the theory of personal constructs from social psychology (Kelly, 1955). It is concerned with developing knowledge about perceived chains of causality and in the development of understanding and knowledge through dialogue about different perceived chains of cause and effect. A key notion drawn from Kelly is that concepts are defined in terms of their polar opposites and therefore a significant location of knowledge is in these polarities.
- *Soft Systems Methodology* (SSM) (Checkland, 1981; Checkland and Scholes, 1990) is based on systems theory. It attempts to describe holons: mental models of systems, not systems in the real world. Knowledge is created through the discussion of these holons and through modelling the consequences of adopting particular interpretations.
- *Strategic Choice* (Friend and Hickling, 1997) proceeds by a codification of facilitative practice. It is a facilitated workshop methodology which develops concepts (categorised as decision, uncertainty and comparison areas) through discussion and attempts to ascribe consensual meanings to these concepts through the agreements of labels for concepts that stand as tokens of the knowledge embedded in the concepts. Uncertainties, particularly those recognised as uncertainties about values, as described in the approach, in part represent ambiguities of meaning and the approach is designed to provoke participants into investigations to reduce these uncertainties.
- *Drama Theory* (Howard et al., 1993) is a radical extension of traditional game theory.
- *Robustness Analysis* (Rosenhead, 2001) proceeds by the application of mathematical modelling approaches to non-numeric data.

The above methods have in common the centrality of the problem situation based on the problem owners' perceptions and interactive processes to enhance their knowledge of the problem as a basis for future action and improvement (Venters, 2001).

5.2.5 *Organisational conversation*

There is wide agreement that communication in its various forms is essential to knowledge nurturing within groups and communities. Conversation takes an important role in such communications. Conversation has been defined as 'a calibration of our own mental models against those of others around us' (Goleman, 1985). In the context of organisations, informal and formal conversation (face-to-face or through document sharing or e-mail communication) is the main method of sharing information and developing common understandings of problem situations: 'Conversations are the way that knowledge workers discover what they know, share it with their colleagues, and in the process create new knowledge' (Webber, 1984).

Organisational conversations take place within formal or informal social structures. It facilitates knowledge sharing and elicitation, and thus knowledge management initiatives should endeavour to exploit and enrich such conversations. In fact, 'Knowledge is not communicated. Knowledge is a critical social product accomplished in communication' (Deetz, 1992).

One form of organisational conversation is the telling of stories. 'Stories infuse events with meaning . . . through the magic of plot' (Gabriel, 2000). Stories are seen as a method by which individuals present events and experience as the storyteller wishes to believe they happened rather than as they actually happened. Their narrative and plot allow them to be remembered by others, altered and shared to allow meaning to diffuse (Gabriel, 2000). Such stories can be used as knowledge-sharing devices, to be created and shared to pass on experience and knowledge (Snowden, 2002). It should however be noted that once an individual has told their story they lose control of the story and its meaning. Its message may be altered and even reversed through subsequent telling (Gabriel, 2000).

5.2.6 *Knowledge mapping*

Effective knowledge management programmes require a holistic understanding of the socio-organisation environment in which the knowledge is used. The political nature of knowledge and its link to human activity requires a method which can address such flexibility, as suggested by Blackler (1995): 'Knowledge is analysed as an active process that is mediated, situated, provisional, pragmatic and contested. The approach suggests that attention should be focused on the systems through which people achieve their knowledge and on the processes through which new knowledge may be generated.'

Knowledge mapping methodologies have been developed within the field of knowledge engineering to codify knowledge. The following principles from knowledge engineering may be of use in knowledge management (Shadbolt and Milton, 1999):

- Recognise that there are different types of knowledge;
- Recognise that there are different types of expert and expertise;
- Recognise that there are different ways of representing knowledge;
- Recognise that there are different ways of using knowledge;
- Use structured methods.

These principles are clearly derived through the functionalist view of knowledge management – considering knowledge within organisations as a resource to be catalogued and stored (Venters, 2001).

5.3 Knowledge creation processes

This section presents different knowledge creation models. The SECI model (Nonaka *et al.*, 2000) is first presented, followed by three related models. A comparative analysis of these models is provided at the end of this section with recommendations related to the nature of the construction sector.

5.3.1 *The SECI model*

The SECI model (Nonaka *et al.*, 2000) perceives knowledge creation as the spiral interaction process of knowledge conversion between tacit and explicit knowledge. The knowledge conversion includes four modes: socialisation, externalisation, combination and internalisation. The socialisation mode highlights the conversion of tacit to new tacit knowledge through shared experience (e.g. apprenticeship). The externalisation mode focuses on the conversion of tacit knowledge to explicit knowledge by creating concepts articulating tacit knowledge (e.g. metaphor, analogy and model). The combination mode refers to the conversion of explicit knowledge to new explicit knowledge that is more systematic. The internalisation mode involves embodying explicit knowledge into tacit knowledge through learning by doing.

Organisations need to establish a place or a space, '*ba*', to create knowledge (Nonaka and Konno, 1998). This is a requisite as knowledge cannot be created without context. '*ba*' is a shared place, including physical or virtual, for creating knowledge through human interaction. Four types of *ba* within the SECI process are identified: originating *ba*, dialoguing *ba*, systemising *ba* and exercising *ba*. Originating *ba* is a common place for sharing experience through face-to-face interactions. Dialoguing *ba* is a place where mental models and skills are articulated by common terms or concepts. Systemising *ba* is a place of collective and virtual interaction, where people can have activities through on-line networks or any computer technologies. Exercising *ba* is the place for embodying explicit knowledge through virtual interaction.

Knowledge assets are the inputs, outputs and moderating factors of the knowledge creating process. They are divided into four types (Nonaka *et al.*, 2000):

(a) experiential knowledge assets, consisting of the shared tacit knowledge built through organisational experiences; (b) conceptual knowledge assets, consisting of explicit knowledge articulated through images, symbols and language; (c) systemic knowledge assets, consisting of systemised and packaged knowledge; and (d) routine knowledge assets, consisting of the tacit knowledge that is routinised and embedded in the actions and practices.

To lead the knowledge-creating process, top and middle managers are identified as the key persons to work on the four elements of the process (Nonaka *et al.*, 2000). They have to provide the knowledge vision, develop and promote sharing of knowledge assets, create and energise *ba*, and continue the spiral of knowledge creation.

5.3.2 7C model

The '7C model' is proposed by Oinas-Kukkonen (2004) for understanding organisational knowledge creation. The 7Cs (which consist of connection, concurrency, comprehension, communication, conceptualisation, collaboration, and collective intelligence) play a critical role in the knowledge creation process. The 7C model is described as the dimension of different contexts: technology, language and organisational contexts (Lyytinen, 1987). The technology context is best illustrated through Internet technology which provides a platform to share and deliver knowledge to users regardless of space and time. In the language context, 'comprehending' and 'communicating' are introduced as important factors through which information is provided to users. In the organisational context, knowledge 'conceptualisation' can articulate knowledge through interaction among people ('collaboration'). This '7C model' leads to a greater sense of togetherness and 'collective intelligence'.

The 7C model is not linear, but a multiple-cycle spiral process (Oinas-Kukkonen, 2004). Four key phases or sub-processes driven within the knowledge creation exercise are proposed: comprehension, communication, conceptualisation and collaboration.

Comprehension refers to a process of surveying and interacting with the external environment and embodying explicit knowledge into tacit knowledge by 'learning by doing' (similar to internalisation in the SECI model). Communication refers to a process of sharing experiences (similar to socialisation in the SECI model). Conceptualisation refers to a collective reflection process articulating tacit knowledge to form explicit concepts and systemising the concepts into a knowledge system (similar to externalisation and combination in the SECI model). Collaboration refers to a true team interaction process of using the produced conceptualisations within teamwork and other organisational processes.

5.3.3 Community-based model

From the models mentioned above, Lee and Cole (2003) proposed an alternative model of knowledge creation, the 'community-based model'. The latter exhibits substantial differences with the SECI model: it does not concentrate on the individual or a firm, while the SECI model does. The community-based model focuses on knowledge creators who are talented volunteers and on interactions across organisational and geographical boundaries. In other words, the knowledge created is owned by anyone who contributes to it creation. Table 2.2 highlights the major differences between the firm-based and the community-based models of knowledge creation.

This model is quite interesting in the context of the construction industry as communities of practitioners originating from different companies collaborate together in delivering services and products. There are inherent informal knowledge creation processes that are worth exploring.

5.3.4 *Combined research model*

To compete effectively in a dynamic global market, there is a need for decision-making environments to assist end-users and managers in their decision making. Heinrichs and Lim (2005) propose the 'combined research model', combining organisational decision models and competitive intelligence tools. Four factors of knowledge creation and strategic use of information competence are identified:

- *Pattern discovery*: pattern discovery drives organisations to create new knowledge from existing knowledge such as past decisions, past solutions and diagnostic evaluation of past rules and models.

Table 5.1 The comparison between the firm-based and the community-based model of knowledge creation (Lee and Cole, 2003)

Organisation principles	The firm-based model	The community-based model
1. Intellectual property ownership	Knowledge is private and owned by the firm.	Knowledge is public but can be owned by members who contribute to it as long as they share it.
2. Membership restriction	Membership is based on selection, so the size of firm is constrained by the number of employees hired.	Membership is open, so the scale of the community is not constrained.
3. Authority and incentives	Members of the firm are employees who receive salaries in exchange for their work.	Members of the community are volunteers who do not receive salaries in exchange for their work.
4. Knowledge distribution across organisational and geographical boundaries	Distribution is limited by the boundary of the firm.	Distribution extends beyond the boundary of the firm.
5. Dominant mode of communications	Face-to-face interaction is the dominant mode of communication	Technology-mediated interaction is the dominant mode of communication.

- *Strategy appraisal*: appraising the impact of a strategy is necessary before deciding to continue or develop new niches, and allows organisations to develop an historical knowledge base regarding the success and failure of past strategic decisions.
- *Solution formulation*: formulated solutions are key components affecting insight generation competence and can gain higher confidence of knowledge workers.
- *Insight generation*: Insight generation involves observing and interpreting charts, graphs, tables and other information to derive meaningful ideas, directions and solutions for the organisation. Insights can provide guidance to innovative problem solving and strategic decision-making.

This model suggests that technology, including databases, data warehouses, knowledge bases and decision support systems environments can play a key role in promoting knowledge creation processes within individuals, teams, and organisations.

5.4 Conclusion

This chapter introduced different perspectives and approaches to knowledge management and has touched on important aspects related to knowledge creation. From the above, it is clear that knowledge cannot exist independently of human experience and social practice, as it is continuously shaped by the social practices of communities of individuals. Several authors have attempted to conceptualise the relationship between tacit and explicit knowledge in knowledge management activities. Tacit knowledge is personal, context-specific and hard to formalise or communicate, whereas explicit knowledge is disseminated in formal language using knowledge-engineering techniques.

While mapping, storing and disseminating knowledge is important, this chapter highlights the importance of understanding the socio-organisational environment in which knowledge activities take place. It is essential to support individuals in making sense and exploiting knowledge.

Sensemaking emphasises the way individuals make sense of their organisational context. Sensemaking suggests that in order to build a form of understanding of knowledge individuals are affected by their own context and are driven by plausibility rather than accuracy. Such complex action suggests that simplistic 'codification' approaches can show limitations in knowledge management activities.

Hence, it is important not to overlook the interdependence between knowledge and the socio-organisational context. This takes an even more important role in the context of the project-based nature of the construction industry which involves several organisations sharing, exploiting and creating knowledge. The community-based model provides some interesting insights that suggest that communities of interest and communities of practice that transcend organisational boundaries are the way forward to knowledge value creation.

6 Knowledge management systems

Knowledge and ICT evolution; Traditional document management systems; Corporate portals and project extranets; Groupware systems; Service-based knowledge solutions for the construction virtual enterprise; Advanced information management environments; Decision support systems; Common and emerging functionality of knowledge management systems; The eCognos knowledge management platform; Text mining techniques; Conclusion.

6.1 Knowledge and ICT evolution

Advances in personal computer technology along with the rapid evolution of graphical user interfaces, networking and communications have had over the last decades a substantial impact on industry business processes. The emergence of client/server applications (ranging from file server to database server applications), at the end of the 1980s, has offered a first promising answer to the problems of scalability of modern businesses (ability to upgrade a system without having to re-design, for instance the ability to extend the underlying data structures). Software applications were being downsized from expensive mainframes to networked personal computers and workstations that are often more user-friendly and cost effective (Rezgui, 2001).

The Internet has brought new challenges that the industry is now trying to comprehend and exploit. On the other hand, efforts are continuously deployed to web-service and integrate legacy and proprietary systems with web-based environments (Rezgui, 2007d). These legacy, proprietary and commercial applications widely used by industry embed useful data, information and knowledge, and tend to be generally known as information systems. They range from in-house document management solutions to complex business-oriented applications.

Previous chapters have made the argument that technology is the cornerstone of the construction project, assimilated to a virtual enterprise, by enabling project actors to work collaboratively although geographically dispersed. Technology not only supports the tasks which individuals and organisations need to perform, but enables knowledge and information to be shared and thus transcends organisational boundaries. New technologies are constantly emerging which offer potential benefits to organisations for improved efficiency and effectiveness, enhanced competitiveness and delivery of new products and services.

A knowledge management system (KMS) refers to a class of information systems applied to managing organisational knowledge (Alavi and Leidner, 2001). They are developed to support and enhance the organisational processes of knowledge sharing, transfer, retrieval and creation. The literature discussing applications of IT to organisational knowledge management initiatives reveals three common applications (Alavi and Leidner, 2001): (a) the coding and sharing of best practices, (b) the creation of corporate knowledge directories, and (c) the creation of knowledge networks. The following sections adopt a different perspective to Alavi and Leidner's common KM applications, and instead review categories of information systems that embed organisational or project knowledge. The chapter discusses these knowledge systems in terms of their functionality and applicability to the construction industry.

6.2 Traditional document management systems

Construction is a knowledge-intensive industry as noted earlier in the book. Most information used during the design and build process of a construction project is conveyed using documents. These are most of the time exchanged, for contractual and legal reasons, on a paper-based medium, even when produced using computers. The challenge that the industry is facing is the mining, codification and re-use of the knowledge and lessons learned stored within these documents. Construction project documents tend to be unstructured in their large majority which cause some problems in terms of managing their internal contents and semantics.

Document management has become a crucial issue within modern construction companies. Many leading construction organisations, with an advanced IT department, have undertaken the development of their own tools and solutions to support the production and maintenance of project documents. Even though such proprietary tools provide many helpful facilities, including support for document storage, retrieval, versioning and approval, they don't handle any semantics of the information being processed and therefore remain limited in their support of end-users' information and knowledge needs. In fact, construction project data (such as IFC) and documentation (including full specification documents) constitute two fragmented information sectors where compatibility and interoperability are mostly needed. Moving these pseudo-sectors closer together to support construction project documentation as part of the lifecycle of the building product is becoming an actual and urgent topic for standards bodies and industry alike.

A number of limitations of traditional electronic document management (EDM) systems can be noted (Hjelt and Bjork, 2006; Rezgui, 2001; Rezgui and Karstila, 1998):

- Every partner within the project must use the same EDM system on a project to access and share documents.

- The document's semantics is not controlled by the EDM system. Documents are handled as 'black-boxes'.
- The EDM system does not support document semantic cross-referencing. This is usually handled by classifying related documents under the same category or folder.
- Security is always an issue. It is not easy to implement as for printed documents. EDM systems require good levels of user authentication and document protection.
- The EDM system is often not integrated with proprietary and commercial business applications used within the company (e.g. CAD applications and word processors).
- A majority of end-users in the construction industry are not computer-literate. EDM systems lacking user-friendliness, or subject to slow network problems, tend to discourage users from their full adoption.

With the advent of the Internet, several EDM systems have been upgraded to become web-enabled. It has helped resolve some of the above problems including the advantage of using a universal browser. In addition, a number of commercial web-based EDM systems have been developed, including ProjectNet (http://www.citadon.com/products/projectnet.htm) and Buzzsaw (http://www.autodesk.com/buzzsaw). These systems provide document and workflow management services across the Internet. However, none of these systems provides a framework for managing the internal semantics of documents in a way that promotes knowledge re-use as described in section 6.8.

6.3 Corporate portals and project extranets

Intranets have attracted a lot of interest from construction companies with a view of managing internal (corporate) documentation. Similarly, extranets have been used to manage documents produced in multi-actor environments, such as construction projects. These are perceived as a logical development and migration of existing traditional document management systems into web-based forms that exploit the ubiquitous dimension of the Internet. Several systems illustrate this evolution as reported in (Nitithamyong and Skibniewski, 2006; Rezgui, 2006). The advantages of such systems include a consistent and structured approach to sharing and updating documents in corporate and multi-actor environments, resulting in an overall improvement in the quality of documents (up-to-date and timely information). From a technical perspective, project intranets/extranet use Internet protocols to manage documents on centralised servers. A universal web browser is used as a gateway to share documents anywhere/anytime, thus addressing problems that usually occur in linear communication approaches. Extranets tend to include useful functionality for searching documents based on key-word or full search mechanisms (e.g. standards, regulations, full specification documents, etc.) or conducting project-related business transactions (e.g. financial project management, electronic bidding and procurement).

This class of information systems is commonly referred to as web-based project management systems (WPMS) (Nitithamyong and Skibniewski, 2006). Currently, the most widely used WPMS are managed by application service providers (ASPs) who host the entire and related services (including document servers) linked to managing projects. Moreover, they provide the computing power, storage, security, backup, network infrastructure and technical staff required to manage the platform. These are adapted to the nature of the construction industry that is dominated by a majority of SMEs that cannot afford to invest in expensive IT facilities. Examples of leading UK-based systems are those provided by the Network for Construction Collaboration Technology Providers (NCCTP) members, including 4Projects; Aconex; ASite; BIW Technologies; Business Collaborator; CadWeb; Causeway; and The-Project. US-based leading systems, for example, include Buzzsaw and Constructware by Autodesk; Sword-CTSpace by Sword Group; ProjectTalk by Meridian Project Systems; PrimeContract by Primavera; and ProjectWise by Bentley. Apart from these leading WPMSs, there are some popular low-cost systems which offer limited WPMS functions for small and medium enterprises, such as Woobius and Collabor8online (Nitithamyong and Skibniewski, 2006).

However, the adoption of such systems in the construction industry is relatively limited as illustrated by a recent survey conducted by NCCTP (2006). It is interesting to note that 44 per cent of the survey respondents were satisfied with their WPMS experience. However, they are unsure about generalising the use of WPMS on every project. Three per cent questioned the need for such environment (NCCTP, 2006).

6.4 Groupware systems

A great deal of research has been done in the field of computer support for co-operative work (CSCW). CSCW is more generally concerned with the introduction and use of groupware systems to enable and support team work. Groupware deals with documents ranging from highly structured to unstructured data, including text, images, graphics, faxes, mail and bulletin board. Groupware solutions include traditionally a subset of the following system components: workflow (task scheduling), multimedia document management, e-mail, conferencing and a shared schedule of appointments.

Groupware provides means of flattening organisations and removing layers of bureaucracy. From a construction project perspective, its functionality has the potential to help manage and track information, documents, users and the applications they use. It also allows actors collaborating on specific tasks to exchange and share information and synchronise their work. It offers the potential to maintain the so-called 'project memory' and record all lessons learnt in a way that promotes re-use.

Workflow is a key functionality of a groupware system. A workflow process consists of a collection of activities. An activity is a logical step that contributes toward the completion of a workflow process. It is executed by a dedicated application software outside the workflow system.

However, there is a strong reluctance in the construction sector to adopt or join a workflow-based process, or to fully use a groupware solution (Nitithamyong and Skibniewski, 2006; Rezgui, 2007d). There are still important requirements to be addressed, including ensuring the availability of up-to-date, accurate and relevant information, while at the same time providing access control (based on actors' rights and role in the project) and concurrent business transaction support. In fact, the peculiarities of the construction sector suggest that equal consideration should be given to technical as well as to social, contractual and legal aspects (including responsibilities) relating to the complex nature of construction projects.

A suitable workflow technological solution has to demonstrate capability of supporting the central business processes, allow integration of systems and interoperability between disparate applications and enable the management of interactions between individuals and teams, whilst at the same time taking into account the fact that the industry is dominated by SMEs (the majority of which do not exceed 19 employees) and operate within tight financial margins.

These issues suggest that construction projects should be considered as Virtual Enterprises (VE) and that, as such, they require solutions adapted to the nature of work and business processes that underpin the design, construction and delivery of a building project.

6.5 Service-based knowledge solutions for the construction virtual enterprise

Calls have been made for technology solutions adapted to the nature of team work and the dynamic networks of the construction industry (Ruikar and Emmitt, 2009). A variety of software applications are used to perform various business processes and activities. However, the ability to integrate disparate, heterogeneous enterprise information systems to implement a distributed business process is a key challenge faced in any virtual enterprise setting.

Service-oriented computing has become the prominent paradigm for leveraging inter and intra enterprise information systems, creating opportunities for smart organisations to provide value added services and products. Web services have emerged as a serious technology to provide the middleware platform to support effectively the operations of a VE. Web services are self-contained, web-enabled applications, capable not only of performing business activities on their own, but also possessing the ability to engage other web services in order to complete higher-order business transactions (Yang, 2003). The benefits of web services include the decoupling of service interfaces from implementation and platform considerations, the support for dynamic service binding, and an increase in cross-language and cross-platform interoperability (Curbera, 2002; Yang, 2003). There is a need now for this new form of computing to move from its initial 'Describe, Publish, Interact' capability, to transforming services into reinvented assemblies in ways that previously could not be predicted in advance (Van Den Heuvel and Maamar, 2003).

Research indicates that an approach that promotes application hosting through service federation as opposed to the traditional 'purchase/build/self-manage' applications is more suited to the nature of construction (Rezgui, 2007a). Moreover, the service-oriented computing paradigm enables the migration of the traditional standalone-hosting model to a networked one, by allowing web services to dynamically discover and hook-up to web services offered by different providers. The service-oriented virtual enterprise allows the creation of domain-specific vertical libraries of services that are modular, well documented, implementation-independent and interoperable (Rezgui, 2007d). This involves federating services from various non-collocated organisations and software houses and making the tools and services they offer available via ubiquitous web browsers. A potential new ICT business model for the construction sector would be that of the 'rental' of services via a platform designed to support a construction VE. This new form of software licensing would provide a software service to include configuration, maintenance, training and access to a help-desk. It would enable organisations, including SMEs, to rent instead of purchase software. This would involve using only the functionality directly needed by users, therefore reducing cost and increasing work efficiency.

This approach will essentially provide a scalable and user-friendly environment to support teamwork in construction by: (a) delivering to clients a customised VE solution maintained by a dedicated application service provider; (b) providing an alternative to the traditional licensing model for software provision by introducing a model based on service rental or offered on a pay-per-use basis; (c) providing a change of focus from 'point to point' application integration to service collaboration and inter-working; (d) delivering higher-order functionality, composed from elementary services, providing direct support for business processes; (e) providing a ubiquitous dimension to business processes, as services can be invoked anytime, anywhere from a simple web-browser and (f) enabling a single point of contact for service and client support (Rezgui, 2007a).

6.6 Advanced information management environments

Product data technology (such as the Industry Foundation Classes) and related semantic resources (taxonomies, thesauri, dictionaries, etc.) provide means of managing information semantics in an integrated way. Managing this information effectively involves addressing a number of central issues (Rezgui *et al.*, 1998), including:

- *Ownership, rights and responsibilities*: each actor is assigned a specific role in the project through which he or she interacts with the project information base. Providing support for actors' rights, ownership and responsibility over information is essential in order to maintain a reasonable level of information consistency.
- *Versioning of information*: it is a mechanism used to keep track of all the states in which an object has existed, including its current state. It can also be

useful to keep alternative versions of current information to deal with situations in which a decision has not yet been made from a number of possible alternatives.

- *Schema evolution:* the lifecycle of a building, from inception to demolition, could span anything from a couple of decades to several centuries. It is important in this context to allow the underlying conceptual model of a building project to be altered, and evolve over time, without affecting the overall consistency of the project information base.
- *Recording of intent behind decisions leading to information:* a great deal of information is created as a result of formal or informal decisions. These decisions are influenced by various factors. It in important to provide an environment for recording the factors that influence a decision that leads to information.
- *Tracking of dependencies between pieces of information:* project information is all inter-linked and dependent on one another. It is equally important to provide an environment where dependencies between versions of information are tracked and stored.
- *Notification and propagation of changes:* it is the process by which actors are kept informed by any change introduced to the project information base, provided that it is relevant to them. For instance, the civil engineer should be informed by most of the amendments made by the architect to the project (adding or removing walls, increasing the height of a storey, etc.).

The most significant issues arising from managing information are concerned with the level of granularity at which various forms of information management should be applied (ranging from documents to data). This issue can be addressed by providing sufficient flexibility in the models to allow for variations in granularity, and by introducing a mechanism for aggregating changes for the purpose of notification and recording of intent.

An illustration of these principles is described in Rezgui *et al.* (1998). These concepts have later been incorporated into CAD environments (Cooper *et al.*, 2005).

6.7 Decision support systems

Knowledge-based systems (KBS) have now been in use for several decades in industry. The UK department of Trade and Industry reported in one of its early surveys over 2,000 knowledge-based systems deployed in business operations in industry back in the early 1990s, including manufacturing. Overall, these systems work fairly well on problem domains that have an explicit model-based representation implemented through rules or objects.

However, developing KBS without an explicit problem domain model remains problematic. That is where other forms of reasoning, including case-based reasoning (CBR), have been explored. Case-based reasoning organises the structured archiving of past experiences for future potential re-use. These experiences, commonly referred to as cases, are archived along with their unique domain characteristics expressed through well-defined indexes that describe the essence of the case (Marir *et al.*, 2000). Several studies and prototype implementations have been proposed which highlight the benefits of decision support systems using CBR techniques (Marir *et al.*, 2000; Watson and Marir, 1994). The success of this type of reasoning is largely explained by its simplicity, along with its similarity with human problem-solving mechanisms. Case-based reasoning, compared to knowledge-based systems, provides many advantages, including the following ones:

- CBR does not require an explicit domain model and so elicitation becomes a task of gathering case histories.
- Implementation is reduced to identifying significant features that describe a case.
- CBR systems can learn by acquiring new knowledge as the number of cases increase.

Most available knowledge management systems (including KBS and CBR) rely on users' input to orchestrate the information and knowledge discovery and elicitation. This is, however, becoming increasingly complex as the electronic sources of knowledge are vast and rapidly growing with the successful deployment of IT systems within organisations. In addition, these systems have limited collaborative functionality and do not encourage information and knowledge discovery – they require the user to have a clear idea of their knowledge requirements. Such systems also require the tacit knowledge giver to be able to clearly articulate their knowledge and experiences. Also, problems as found in the architecture and engineering disciplines can be very complex and require an adapted form of evolutionary algorithms. This forms the focus of Chapters 9 and 10 of the book.

6.8 Common and emerging functionality of knowledge management systems

A diversity of proprietary and commercial knowledge management environments is used in industry. The selection of the appropriate environment depends on a number of factors, including organisational culture, employees' technology maturity, compatibility with existing systems, and budgets (Nitithamyong and Skibniewski, 2006). Knowledge management systems exhibit a range of functional features in common, including:

- *Search facility:* This is an important aspect of a KMS as the role of such system starts with helping users search and locate a particular expertise, document or knowledge. Search engines have different capabilities and may vary in the way they address users' queries. Ways in which they proceed include the following:
    - *Full text search:* the search is based on a keyword, combination of keywords, or sentence and proceeds by searching all target document

repository or Web about the occurrence of such keyword(s) or sentence. Full text search is provided by most search engines.

- *Semantic search:* the search proceeds by extending the search to concepts related to the ones provided by the user in the form of a sentence or a set of keywords. This assumes that the search engine makes use of a semantic resource (taxonomy or ontology).
- *Profiles or context-based search:* this involves factoring in the user profile (information interests/subscriptions, job description, background, etc.) into the search in order to retrieve knowledge directly related to the user.
- *Collaborative search:* users have the possibility of turning a refined search result into a 'knowledge' interest that is stored as part of the KMS knowledge-base and that can then be re-used by other users with similar or related queries. It assumes that users who share similar interest will find similar documents.

- *Discussion forum:* it provides a place for users to share concerns, questions, answers, and any other experiences. Discussions can take place in a synchronous or asynchronous form. They can be threaded and filtered according the user's profile. Discussion forums are effective means to set up communities of interests.
- *Document review:* This functionality supports document history/version control in access/rights controlled multi-user environments. Documents can be commented and annotated.
- *User profiling:* users register to the KMS and provide information related to them in terms of job specification, discipline, information interests, etc. The profile is managed dynamically and evolves as the user interacts with the system through query relevance feedback.
- *Workflow functionality:* this provides means of controlling and orchestrating the execution of a task, activity or business process involving several software applications and users.
- *Auto finding experts:* This functionality helps locate users based on their expertise or involvement in a given task, activity or project.
- *Taxonomy management:* this functionality allows administrators to manage knowledge categories or taxonomies used to organise knowledge, including documents, and search and structure responses to user queries.
- *Knowledge-base generation and management:* this functionality is used to generate a semantic representation or indexing of documents managed by the KMS. For instance, each document is mapped on to a semantic vector composed of index terms/ontology concepts and a weighting representing the importance of the term/concept in the document.

A plethora of commercial document management environments have been developed which implement a sub-set of the above functionality. These include:

- IBM Lotus QuickPlace (www.lotus.com/quickplace) is one of IBM's Lotus Domino family software. It is a groupware environment adapted for knowledge management purposes.
- SilentOne (www.silentone.com) is a document management system that uses the concept of a knowledge library. It provides a centralised knowledge repository to store and share knowledge supported by a search engine and user authorisation features.
- AskMe (http://www.askmecorp.com) promotes knowledge-sharing networks.
- Tacit ActiveNet (http://www.tacit.com) supports collaborative problem solving.
- 80–20 software (http://www.80-20.com) is a web-based, document management software. It has the function of document version control, full content search and retrieval, role-based access control, save search, etc.
- ProjectNet (http://www.citadon.com/products/projectnet.htm) is a web-based document and workflow management for the design, engineering and construction industry. It belongs to a family of web project management systems.
- Bricsnet (http://www.bricsnet.com) is designed to support project management. Activities. It includes modules to manage workflow, documents and project finances.
- Autodesk Buzzsaw (http://www.autodesk.com/buzzsaw) is an on-demand collaborative project management solution for all industries. It enables team workers to communicate, share, and track project information across all teams in any industries such as architecture and engineering, construction, education, manufacturing, real estate development, etc.

Also, several organisations have developed their own in-house knowledge management solution. The following section gives an overview of the eCognos platform, which is one of the first research-led KM solutions for the construction industry, involving leading European construction organisations (Rezgui, 2006).

6.9 The eCognos knowledge management platform

The eCognos project aims at developing model-based adaptive mechanisms that can organise documents ranging from unstructured (black-box) to highly structured (e.g. XML) ones according to their contents and interdependencies. This relies on an ontology (extended version of the IFCs) which serves as a basis for knowledge indexing, discovery and retrieval using information retrieval (IR) techniques.

The services provided by the eCognos platform include (Figure 6.1):

- *The Ontology Service*: the Ontology service provides the functionality required to make the eCognos ontology available to the other eCognos services which may require it. This is achieved via the ECognos Ontology Server that will be devoted to handling all the ontology-related queries within the eCognos platform.

- *The Visualisation Service*: this offers means for eCognos users to visualise documents according to the type of device in use (e.g. PDA, laptop with low resolution, or laptop/desktop with high resolution) and their own display preferences.

- *The Maintenance Service*: the purpose of the knowledge maintenance package is to enable the system to maintain consistency amongst the knowledge items represented in the eCognos system. The API (Application Programming Interface) presented by this package consists of functions for the specification and enabling of maintenance tasks.

- *The Knowledge Indexer Service*: it provides a means to produce ontological full text indexing of each document managed by the eCognos platform. The indexed version of the document is then used to perform possible initial searches or submit the document to produce an ontology-based semantic representation.

- *The Extractor Service*: this package presents functions that enable the user to perform knowledge extraction tasks. This consists of extracting a representation of a knowledge item in terms of ontological concepts. The result of this extraction will be to make the knowledge item available for manipulation within the eCognos system via its ontological representation. It will also make the item available for indexing and consistency checking. Its API will expose functions enabling the extraction of knowledge representations from knowledge items.

- *The Knowledge Searcher Service*: this package provides functionality enabling the user to perform searches across knowledge items or documents represented in the eCognos system, and to allow them to retrieve those knowledge items and their representations from the system. This service has the capability to enhance and expand a user query with related concepts from the ontology to improve the precision and recall of the returned document set.

- *The Discoverer Service*: the purpose of the eCognos knowledge Discoverer package is to provide functionality that will allow the user to search for useful documents and make them available to the system. This will be knowledge that was not previously represented in, or known about by, the eCognos system. For example, the process of knowledge discovery could be brought about by searching the Internet for knowledge not previously known about by either the user or the eCognos system. These may typically represent documents managed by other related portals at a national or international level.

- *The Disseminator Service*: The purpose of this interface is to provide methods, which enable the dissemination of knowledge and documents to other users of the system through a 'push' mechanism. This includes the flagging of knowledge representations methods enabling the messaging between knowledge representations.

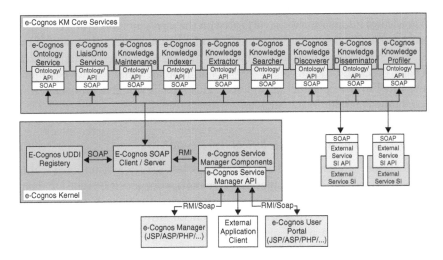

Figure 6.1 eCognos service architecture (adapted from Rezgui, 2006).

- *The Profiler Service*: the purpose of this service is to perform and manage user profiles to better direct user searches.

A comprehensive validation of the platform has been performed. The validation results suggest that this type of environment can support knowledge sharing and creation at three levels:

- *Organisational level*: eCognos is perceived as having the potential to leverage intellectual capital both inside and outside the organisation. Its 'experience sharing' service contributes to maintain the integrity of the social communities in which knowledge is embedded.
- *Team level*: Knowledge sharing is perceived as possible across distance and users can easily access others' experiences. Teams on projects have the potential to combine distributed competences more effectively. Social relationships between team members and communities are nurtured rather than focusing on knowledge storage and retrieval activities.
- *User level*: eCognos addresses the actual needs and requirements of end-user to share knowledge, rather than those of management who wish to control it.

Techniques employed by the eCognos services made use of information retrieval (IR) sciences. An overview of these techniques is given in the following sections.

6.10 Text mining techniques

Explicit knowledge is traditionally conveyed through text. As such documents are an important source of information and knowledge but require human operators to decode and interpret the meaning of embedded information and knowledge. Information retrieval deals with the representation, storage, organisation of and access to information. The information retrieval community has now been working for decades on means of making sense of textual information (Baeza-Yates and Ribeiro-Neto, 1999). This is an area that has not been fully exploited in the construction industry. This section will give an overview of established text mining techniques.

6.10.1 Documents and their logical representation

A document is a transitional and changing object written and authored within a precise stage of the project lifecycle. Generally, a document is related to many elaborated documents of the project documentary database. A document has one or many authors. It is described by general attributes such as a code, an index, a designation, a date of creation and a list of its authors. Ideally, a list of document versions also keeps track of any amendments made to the document during its lifecycle. An indexing system may be associated to the document. A document is submitted for approval according to a defined circuit of examiners representing diverse technical or legal entities. Each examiner issues a statement that enables the document to be approved, rejected or approved under reservation. Also, documents have been traditionally represented using a set of key words. These key words or indices can either be manually defined by a user with a good knowledge of the semantics of the document, or extracted automatically from the text of the document using proven information retrieval (IR) techniques. A document has a logical and a physical structure, which are both used to convey in the best possible way its internal semantics. The physical structure of a document is described using a properly defined syntax supported by one or several software tools.

Each document should have ideally meta-data attached to it. A possible solution for describing meta-data is through RDF (resource description framework – a development based on XML) that provides with a simple common model for describing meta-data on the Web. It consists of a description of nodes and attached attribute/value pairs. Nodes represent any web resource, i.e. uniform resource identifier (URI), which includes URL (uniform resource locator). Attributes are properties of nodes and their values are text strings or other nodes.

6.10.2 A document type taxonomy

Following the description of what a document is, as well as the leading metalanguage and language standards in this area, an attempt is made to classify documents based on their inherent nature and the structure they exhibit, taking into account the specificities of the construction sector. Three classes of documents have been identified, namely: poorly structured documents, documents with a clear physical structure, and highly structured documents.

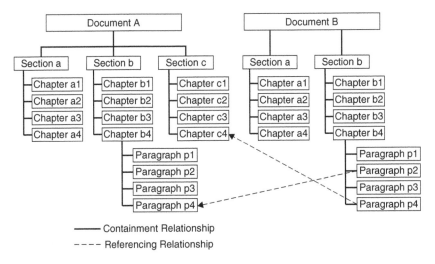

Figure 6.2 Relationship types in a structured hypertext document.

- *Poorly structured documents*: These are documents that are composed of text with no formal structure. These constitute the vast majority of the construction documentation. Documents are treated here simply as black-boxes.
- *Documents with a text formatting structure*: These are documents that are tagged using the HTML language, or at best the XML language but without reference to a (semantic) document type definition (DTD). A physical structure in the form of a hierarchical tree or hypertext link of nodes can be easily generated from this representation as illustrated in Figure 6.2. Each node is associated with a block of text that can represent a paragraph, chapter, section or even a web page. Two related nodes are connected one to the other by a direct link which correlates the text associated with these two nodes. This is explicitly described in the text including by a special tag or highlighted portion of the text. This structure offers a variety of possibilities in terms of text retrieval. These documents include direct references to other documents/document sections.
- *Highly structured documents*: This categorises documents that are instances of an XML-based meta-language. These documents have a semantic structure that can easily be used as a basis for text queries and retrieval. Ideally, we can envisage that all the documentation that is used and produced in the construction industry be an instance of a specific XML DTD over which users can exercise semantics control. These documents include naturally direct references to other documents/document sections.

6.10.3 *Text operations*

This section describes basic text operations that aim at pre-processing documents. There are mainly five text operations: Lexical analysis, Stop words elimination, Stemming, Index term selection, and term categorisation (Baeza-Yates and Ribeiro-Neto, 1999).

6.10.3.1 *Lexical analysis*

The purpose of text lexical analysis is to retain a subset of the initial text in the form of mean full words that are potential candidates for index terms. This involves as highlighted in giving particular attention to hyphens, punctuation marks, digits and the case of letters. In particular digits are not usually good index terms, and they should better be tackled through more conventional solutions like databases. Eliminating hyphens helps treat similar words equally (e.g. 'state-of-the-art' and 'state of the art'). The elimination of punctuation marks shouldn't pose any particular problem, especially within our construction domain. Finally, the case of letters should be all reduced to lower or upper case. Unlike the case of programming languages, this should not have any consequence on the text.

6.10.3.2 *Stopwords elimination*

Stopwords generally do not constitute good discriminators for text retrieval. They include articles, prepositions and conjunctions. It is also arguable that stopwords should also include some verbs, adverbs and adjectives as they provide little added value to the semantics of the text. Documents, through the elimination of stopwords, can therefore be reduced in size, and hence allow better performance in document retrieval and processing.

6.10.3.3 *Stemming*

A stem is a portion of a word that is left after the removal of its affixes (prefixes and suffixes). An example of a stem is *manage* which is the stem for the variants *managing, manageable, manager and management*. Stems are considered to be useful for text processing as they reduce variants of the same root word to a common concept. It has also the advantage of reducing the size of the indexing structure as the number of index terms is reduced. However, there are disadvantages to stemming (Baeza-Yates and Ribeiro-Neto, 1999). As a result of this many web search engines do not adopt stemming techniques.

6.10.3.4 *Index term selection*

Index term selection involves determining which words or combination of words can be used as indexing elements. The decision involves domain experts and may involve semantic as well as syntactic considerations. Semantically, the meaning that an indexing element conveys and its discriminating power are important aspects of an index term. Syntactically, there is an argument that noun words carry more semantics than adjectives, adverbs and verbs (Baeza-Yates and Ribeiro-Neto, 1999).

6.10.3.5 Term categorisation

This involves relating terms to a structure such as a thesaurus with a view, for instance, to expanding an original query with related terms. A thesaurus does not simply involve terms and their synonyms but may include phrases which mean that concepts more complex than simple terms are taken into account and provided within their context. A typical example is *Roget's Thesaurus* which has quite a generic nature. Conversely, there are more specific thesauri such as the ones used in the engineering and scientific communities. Thesauri can be very useful as will be discussed later in this chapter.

6.10.4 Models for document semantics characterisation

Index terms are traditionally used to characterise and describe the semantics of a document. This approach attempts to summarise a whole document with a set of terms that are relevant in the context of the document. While this approach has given some satisfactory results in the area of information retrieval (IR), it still has some limitations as it proceeds by oversimplifying the summarisation process by relying on a subset of relevant terms that occur in a document and uses these as a means to convey the semantics of the document. This section describes the three classical models of IR (Baeza-Yates and Ribeiro-Neto, 1999): Boolean, vector and probabilistic.

In the Boolean model documents are represented as a set of index terms. This model is said to be set theoretic (Gudivada *et al.*, 1997). In the vector model documents are represented as vectors in a t-dimensional space. The model is therefore said to be algebraic. In the probabilistic model, the modelling of documents is based on probability theory. The model is therefore said to be probabilistic. Alternative models that extend some of these classical models have been developed recently. The fuzzy and the extended Boolean model have been proposed as alternatives to the set theoretic model. The generalised vector, the latent semantic indexing, and the neural network models have been proposed as alternatives to the algebraic model. The inference network, and the belief network models have been proposed as an alternative to the probabilistic model. It is also worth mentioning that models that reference the structure, as opposed to the text, of a document do exist. Two models have emerged in this area: the non-overlapping lists model and the proximal node model. These refer to models that combine information on text content with information on the physical structure of the document. A comprehensive survey of structured models can be found in (Baeza-Yates and Ribeiro-Neto, 1999).

6.10.4.1 The Boolean model

The Boolean model is based on set theory and Boolean algebra. Query expressions are provided as a combination of Boolean expressions, including Boolean operators which have clear semantics. It was adopted and had great success in bibliographic and library information systems. The main criticism of the Boolean model (Verhoeff *et al.*, 1961) lies in its binary evaluation system. A document can be either relevant or not to a given query. There is no inherent ability to rank the document in relation to its relevance to a given query. In other words, there is no notion of partial match to the query conditions. It is commonly acknowledged today that index term weighting provides more satisfactory results in retrieval performance. More information on the Boolean model can be found in (Baeza-Yates and Ribeiro-Neto, 1999; Verhoeff *et al.*, 1961; Wartick, 1992).

Alternatives to the Boolean model have been proposed. These include the fuzzy set model and extended Boolean model. In terms of fuzzy set model, the model from Ogawa, Morita and Kobayashi (Ogawa *et al.*, 1991) deserves particular attention in that an ontology is used in conjunction with fuzzy set theory to expand the set of index terms in a query and extend the retrieved document set.

The principle behind the extended Boolean model is to overcome the binary limitations of the Boolean model by extending the latter and enhancing it with partial matching and term weighting from the vector model. This model has been introduced by Salton, Fox and Wu (1983). More thorough description can be found in (Baeza-Yates and Ribeiro-Neto, 1999; Salton *et al.*, 1983).

6.10.4.2 The vector model

The vector model addresses the limitations of the Boolean model by providing an approach that supports document partial matching to a given query. This is achieved by assigning non-binary weights to index terms in documents and queries. These term key word weights are then used in a second stage to sort documents by their level of relevance to the initial query. The vector model is today considered as the most popular IR model (Baeza-Yates and Ribeiro-Neto, 1999; Salton and Lesk, 1968; Salton and Yang, 1973).

Alternative algebraic models include the generalised vector space model, the latent semantic indexing model, and the neural network model. The generalised vector space model assumes that two index term vectors might be non-orthogonal which means that there is a possibility for two index terms to be correlated. This term correlation is used as a basis for improving retrieval performance (Wong *et al.*, 1985).

The principle behind the latent semantic indexing model is that ideas in a text are more related to the concepts that are conveyed within it as opposed to index terms. By using this approach, a document may be retrieved only by the virtue that it shares concepts with another document that is relevant to a given query. As indicated in (Furnas *et al.*, 1988), the intent behind the latent semantic indexing model is to map each document and query vector into *a lower dimensional space which is associated with concepts*. This is achieved by mapping the index term vector into this lower dimensional space (Baeza-Yates and Ribeiro-Neto, 1999).

The neural network model is based on research carried out in the area of neural networks. The principle behind ranking documents that are retrieved against a given query is to match the query index terms against the Document index terms. Since neural networks have been extensively used for pattern-matching purposes, they have been used naturally as an alternative model for information retrieval (Baeza-Yates and Ribeiro-Neto, 1999). Detailed description of this model can be found in (Wilkinson and Hingston, 1991).

6.10.4.3 The probabilistic model

This was introduced initially by Robertson and Sparck Jones (1976) as a mean to address the information retrieval problem within a probabilistic context. It proceeds by refining recursively a guessed initial set of documents matching a user query by involving the user feedback to evaluate the relevance of the retained set. For each iteration, the user retains the documents that best match the query. The system then uses this information to refine the ideal response set. As highlighted in (Baeza-Yates and Ribeiro-Neto, 1999), the main advantage of the probabilistic model is that documents are ranked in decreasing order of their probability of being relevant. The disadvantages include: (1) it is difficult to guess the initial separation of documents into relevant and non-relevant sets, (2) the method does not take into account the frequency in which an index term appears within a document. A thorough description of the Probabilistic model can be found in (Rijsbergen, 1979).

The use of probability theory for quantifying document relevance has always been a field of research in information retrieval sciences. Two examples of IR models based on probability theory are the inference network model and the belief network. Both models are based on the Bayesian belief networks that provides a formalism combining distinct sources of evidence, including past queries and past feedback cycles. This combination is used to improve retrieval performance of documents (Turtle and Croft, 1991).

The inference network model takes an epistemological as opposed to a frequentist view of the information retrieval problem (Turtle and Croft, 1990). It proceeds, as described in (Baeza-Yates and Ribeiro-Neto, 1999), by associating random variables with the index terms, the documents and the user queries. A random variable associated with a user document denotes the event of observing that document. This document observation asserts a belief upon the random variables associated with its index terms. Both index terms and documents are represented as nodes in the network. Edges are drawn from a node describing a document to its term nodes to indicate that the observation of the document yields improved belief on its term nodes. The random number associated with the user query models the fact that the information request specified in the query has been met. This random number is also represented by a node in the network. The belief in the query node is then expressed as a function of the beliefs of the nodes associated with the query terms. A description of this model can be found in (Turtle and Croft, 1990, 1991).

The belief network generalises the inference network model. It was introduced by (Baeza-Yates and Ribeiro-Neto, 1999). It is also based on an epistemological interpretation of probabilities. It differs from the inference network model in that it adopts a clearly defined sample space. It therefore provides a separation between the document and query portions of the network. This has the advantage of facilitating the modelling of additional evidential sources, including past queries and past relevance information.

6.11 Conclusion

The chapter introduces a wide range of technologies and approaches relevant to the construction industry in the area of information and knowledge management.

Document management systems are gaining wide acceptance in the construction industry, and constitute a fairly basic but promising approach to managing explicit knowledge. However, while a plethora of commercial implementations are today available, there seems to be a lack of general adoption on projects.

In terms of product data technology, available standards (IFCs) normalise product information for the purpose of data exchange and sharing. This provides a valuable foundation for product information and knowledge capture, structure, share and re-use. Moreover, product data technology can pave the way for advanced information management to support knowledge elicitation by tracking dependencies between information to infer rational behind modifications introduced throughout the project lifecycle.

In terms of groupware technologies, despite the availability of several commercial offers, construction project actors do not seem to be yet ready to embrace workflow-based processes due to the project-oriented, fragmented and complex (multi-actor and discipline) nature of projects. Instead, service-based approaches that factor in the virtual enterprise dimension of a project seem more promising.

In terms of decision support systems, a plethora of KBS and CBR prototypes/systems have been developed. While these systems provide satisfactory solutions for specialised domain problems, they seldom remain satisfactory when facing the increasingly complex variety of sources of knowledge and the format and medium is which they are stored.

Finally, the potential of text and semantic mining techniques, centred on an ontology, have been discussed. This forms the focus of the following two chapters.

7 Domain conceptualisation through ontology

Semantic resources in the construction sector; From simple product data to complex building information models; Philosophical underpinnings of product data and ontology approaches; Alleviating product data shortcomings through ontology; Reasons behind the low adoption of IFCs; Requirements for a successful ontology; Ontology development methodologies; Conclusion.

7.1 Semantic resources in the construction sector

A variety of semantic resources have been developed in the construction sector ranging from domain dictionaries to specialised taxonomies (as mentioned in section 4.3). The most notable efforts include the following:

- The BS 6100 (Glossary of Building and Civil Engineering terms) (BS 6100, 1992), produced by the British Standards Institution (BSI, an independent national body responsible for preparing British Standards). This is a rich and complete glossary that provides a comprehensive number of synonyms per term that can contribute towards any ontology development effort in the sector.
- The *bcXML* (eConstruct, 2001) is an XML vocabulary developed by the eConstruct IST project for the construction industry. Through bcXML, eConstruct has enabled the creation of 'requirements messages' that can be interpreted by computer applications to locate products and services that meet those requirements.
- The ISO 12006–2 (ISO 12006–2, 2001) is concerned with current classification needs and builds on the experience of developing and using conventional classification systems.
- The IFC (IFC, 2010) (Industrial Foundation Classes) model provides a specification of data structures supporting an electronic project model enabling data sharing across software applications.
- The OmniClass (http://www.occsnet.org/) Construction Classification System (OCCS), developed in Canada by the Construction Specifications Institute, addresses the construction industry's information management needs through a coordinated classification system.

While the semantic resources mentioned above represent the most notable and widely acknowledged efforts in the construction field, Table 7.1 gives a wider snapshot of related semantic resources used today in Europe (Barresi *et al.*, 2005). These include: classification systems and their associated product libraries (used to define and classify construction product components), taxonomies (providing hierarchical conceptualisation of construction disciplines), and various dictionaries/glossaries providing comprehensive definitions of construction terms and concepts.

The IFCs constitute the ideal semantic resource candidate for architecture and engineering practices as explained later in the chapter.

The architecture of the IFC model comprises four layers (IFC, 2010), namely the resource layer, the core layer, the interoperability layer, and the domain layer. The resource layer includes low-level concepts or objects that are independent of both application and domain needs but rely on other classes in the model for their existence. The core layer includes two levels of generalisation, namely kernel and extensions. The kernel provides the high level concepts required within the scope of the current IFC release. It can be seen as a 'template' that defines the form in which all other schema within the model is developed. The core extensions provide a kind of specialisation for the concepts defined in the kernel. In other words, each core extension is a specialisation of a class defined in the kernel. The interoperability layer provides the schemata defining concepts common to two or more domain models. Finally, the domain layer supports the creation of domain models (e.g. architecture, structure, and quantity surveying).

7.2 From simple product data to complex building information models

Although the manual referencing of paper-based product data and building design has existed for centuries, it was the increasing use of CAD facilities in design offices from the early 1980s which prompted the first efforts in electronic integration and sharing of building information and data (Boddy *et al.*, 2007). Here, the ability to share design data and drawings electronically through either proprietary drawing formats or via later de facto standards such as DXF (Drawing/Data Exchange Format), together with the added dimension of drawing layering had substantial impacts on business processes and workflows in the construction industry (Eastman, 1992). Although in these early efforts sharing and integration was mainly limited to geometrical information (Brown *et al.*, 1996), effectively the use of CAD files was evolving towards communicating information about a building in ways that a manually draughted or plotted drawing could not.

This evolution continued with the introduction of object-oriented CAD in the early 1990s by companies such as AutoDesk, GraphiSoft, Bentley systems etc. Data 'objects' in these systems (doors, walls, windows, roofs, plant and equipment etc.) stored non-graphical data about a building and the third party components which it comprises as 'product data', in a logical structure together

Table 7.1 Key semantic resources in Europe

	Uniclass	BS6100	Lexicon	BC Building Definitions	BARBi	SDC	Edibatec
Type	Classification system	Glossary	Vocabulary of terms	Taxonomy	Data library	Dictionary	Dictionary
Access	Free	Online access available to subscribers	Free	GNU Public License	Free access after registration	Free	Free
Location	Thesaurus.for aec/data/ uniclass/ or RIBA Bookshop	bsonline.techin dex.co.uk	www.stabu-lexicon.com	www.bcxml.org	Edmserver. epmtech.jotne. com/barbi/ index.jsp	www.sdc.biz	www. edibatec. org/accueil/ default4.htm
Developed by	National Building Specification Services	British Standards Institution	STABU Foundation	IST e-Construct project	Norwegian Building Research Institute	Gencod EAN France	Edibatec
Published	1997	Various depending on Parts	1997	2002	2003	2002	2000
Scope	Architecture and civil engineering	Building and construction	Building and construction	Building and construction	Building and construction	Building and construction	HVAC
Application	Product libraries	Terms definition	Product catalogues	e-Procurement	Electronic commerce	e-Commerce	e-Catalogues

with the graphical representation of the building (Daniell and Director, 1989; Eastman *et al.*, 2004). These systems often supported geometrical modelling of the building in three dimensions, which helped to automate many of the draughting tasks required to produce engineering drawings.

When combined with the increasing ubiquity of electronic networking and the Internet, this allowed many companies to collaborate and share building information and data which in turn lead to new ways of communicating and working (Bosch *et al.*, 1991; Daniell and Director, 1989). The opportunities presented by the move towards collaborative working and information sharing encouraged a number of research projects in the early 1990s, which aimed to facilitate and provide frameworks to encourage the migration from document-centred approaches towards model-based, integrated systems: CONDOR (Rezgui and Cooper, 1998) and COMMIT (Rezgui *et al.*, 1998) being examples. Similarly, the OSMOS (Rezgui 2007a) research project aimed to develop a technical infrastructure which empowered the construction industry to move towards a computer-integrated approach.

It became clear that in order to take best advantage of the potential for CAD and object/product model integration, there was a need for more coordinated standards which would simplify and encourage its uptake (Gu and Chan, 1995). These standards-defining efforts came in the form of the STEP application protocols for construction (Mannisto *et al.*, 1998). This work, inspired by previous work primarily in aerospace and automotive fields, formed part of ISO 10303, the International Standard for the Exchange of Product Model Data. Latterly, the International Alliance for Interoperability defined the Industry Foundation Classes, a set of model constructs for the description of building elements. Preceding and in some cases concurrent with this work, the research community produced several integrated model definitions including the AEC Building Systems Model (Turner, 1988), ATLAS (Bohms *et al.*, 1994), the RATAS Model (Bjork, 1994), and the COMBINE Integrated Data Model (Augenbroe, 1994, 1995). These research efforts tended to propose a data model and also provide a suite of tools to manipulate the model (as a proof of concept), or a central database to serve model elements to other applications used in the construction project process via some form of adapter (Björk, 1998; Björk and Penttilä, 1998; Giannini *et al.*, 2002).

Within the last three to four years, researchers and commercial application developers in the construction domain have started to develop tools to manipulate complex building models (Mannisto *et al.*, 1998). By storing and managing building information as databases, building information modelling (BIM) solutions can capture, manage, and present data in ways that are appropriate for the user of that data. Because the information is stored in a logically centralised database, any changes in building information data can be logically propagated and managed by software throughout the project lifecycle (Rezgui *et al.*, 1998). Building information modelling solutions add the management of relationships between building components beyond the object-level information in object-oriented CAD solutions. This allows information about design intent to be captured in the design process. The building information model contains not only a list of building components and locations but also the relationships that are intended between those objects (Lima *et al.*, 2005).

This new wave of BIM applications embodies much of the vision of previous academic research such as ATLAS and COMBINE, whilst still relying on data exchange standards or API-level customisation for interoperability/integration. Recently, the American National Institute of Building Sciences has inaugurated a committee to look into creating a standard for lifecycle data modelling under the BIM banner (NIBS, 2007). The idea here is to have a standard that identifies data requirements at different lifecycle stages in order to allow a more intelligent exchange of data between BIM-enabled applications.

7.3 Philosophical underpinnings of product data and ontology approaches

Gruber (1995) defines an ontology as 'a formal, explicit specification of a shared conceptualisation'. Conceptualisation refers to an abstract model of some phenomenon in the world which identifies the relevant concepts of that phenomenon. Explicit means that the types of concepts used and the constraints on their use are explicitly defined. Formal refers to the fact that the ontology should be machine processable (Studer *et al.*, 1998).

Practitioners in the construction industry can find it difficult to see clearly how an 'ontology' differs from what they already recognise as a 'data model', focusing on the formal nature and structuring mechanisms that seem to be characteristic of both. Certainly, data modelling languages provide the ability to define taxonomies through notations that support classification, generalisation and specialisation, they support the definition of relationships or associations between concepts, and ideas of aggregation and composition, and in terms of these primitives appear to offer the same support for representing concepts and the relationships between them.

However, trying to understand the distinctions in terms of the modelling primitives that are used is not appropriate; it is the nature of the models themselves, the way in which they are derived, and the tools that support their use that provides the differentiation. In order to understand this, it is necessary to return to the underlying problems that make it difficult to achieve a single agreed data model for an industry.

Returning to Gruber's definition of an ontology, a key element is the idea of a *shared conceptualisation* (Gruber, 1995). Typically, in human endeavour, shared conceptualisations are defined over a lengthy period of time, based on the shared experience of a group of people, sometimes referred to as a community of practice (Wenger *et al.*, 2002). They will involve the definition and use of abstractions that are designed to capture the important aspects of some practical context in order to support a particular activity or type of activity. As such, a shared conceptualisation is a socially constructed model or reality that is distinct from reality and is optimised to support the goals and activities of the community of practice in which it was defined. Communities engaged in different activities are likely to form shared conceptualisations that are quite different views of reality, and make up shared 'world-views' (Checkland and Holwell, 1998) that provide a basis for highly effective and efficient communications within the respective communities.

In order to understand and formalise the shared world-views of such communities in the form of ontologies to support the integration of diverse human activities, it is important to consider approaches that derive from an interpretive philosophical standpoint rather than from a positivist, scientific/engineering one (Rezgui *et al.*, 2009). In such an approach, we try to interpret, accommodate and model what is, rather than trying to change reality to fit a single model. This inevitably results in different ontologies for different communities, but the challenge then is to find ways to allow those communities to collaborate effectively with one another whilst maintaining their existing, efficient, effective separate world-views. The implication is that we are forced to shift our emphasis from developing a standard representation of a single 'reality', towards providing mechanisms for supporting communication between differing perceptions of reality, focusing our attention on the overlaps at the boundaries and the specific conceptualisations that are required for such communication to happen.

An important consequence of this shift is that it becomes possible to adopt a more incremental approach to the integration of processes across disciplines. The single data model approach can tend to result in the need for an 'all-or-nothing' approach to implementation, and certainly practical issues have been noted regarding the size and manageability of IFC models (Bazjanac, 2004). Even with the existence of product model servers (Eastman, 1999), the practical implementation of single-model-based integration seems fraught with difficulty.

7.4   Alleviating product data shortcomings through ontology

The progress made so far in arriving at the BIM concept and its associated tools is undoubtedly a sizeable step forward in the management, communication and leveraging of construction project information. Both the BIM models used by the commercial vendors and the international standards developed for construction such as STEP and IFC do however still exhibit shortcomings. These are identified in Table 7.2 along with the level of support offered by both the existing popular standards (including IFCs) and the potential ontology based solutions.

It is clear then that the ontology approach is by no means a cure-all for the ills of product data models as they currently stand. Indeed we would not propose to replace data models so much as enhance them, and the applications used to create and manipulate them, with additional semantic information based on domain and core ontologies, as described in the following chapter. However, we would maintain that ontologies have the right interpretive philosophical underpinning that is more likely to address the information and knowledge-sharing requirements of the construction user community.

Table 7.2 Product data versus ontology

Issue	STEP/IFC	Ontology
Information schema evolution through time to support changing building and project descriptions or industry contexts	Not fully catered for. The IFC Property Set construct could be employed to fulfil role for certain types of information.	This is an inherent attribute of ontology although not catered for specifically; but a well-maintained and updated ontology should evolve with the domain quite naturally (whilst respecting and allowing for business processes which may require a stable schema interpretation).
Views on data aligned to user and application needs	Views on STEP models can be defined in Express-X. The application protocols are themselves domain-specific views to some extent. Other work is ongoing to extend models into specific sub-domains.	Base level domain ontologies could be said to be views in their own right as they support the information needs of communities of practice. These may be able to be transformed through the mechanism of an upper ontology to suit.
Object ownership and rights management	Not supported, but can be via EXPRESS	Not supported but can be via OWL
Lifecycle management and placement of data in the process	STEP defines basic resources dedicated to relating data to its place in the process	An ontology of construction would include concepts describing the construction process and would therefore likely feature relations to classes of data involved in the process. This would however be a less fixed notion than the STEP approach.
Recording/ embedding of design decision rationale	Some basic support (Kim et al., 2008)	The semantic-rich description that characterises ontologies provides useful means to record and embed design decision rationale.
Links to external information – particularly unstructured information	Links can be manually defined, but they have no specific semantics.	Ontologies tend to be built from a variety of semantic resources, including text documents using information retrieval techniques. The latter are used to infer external information's 'relation' to ontology concepts and by examining other existing relations, its links to other data.

The Web Ontology Language (OWL) is the current leading international standard notation for the definition of ontologies in a machine-interpretable fashion. OWL has two primary constructs, namely classes and properties. Classes represent categories of things, real or conceptual, and properties define the relationships between classes and between instances of those classes. An OWL class is somewhat analogous to an Express application object and whereas Express really only has explicit relationships of the 'is a' inheritance hierarchy type, OWL properties are far more flexible and explicit in describing a richer set of possible relations between its classes. Many Express relationships are opaquely embedded in the properties of application objects. Similarly, the Express-based IFCs define a slightly broader range of relationships but still somewhat fewer than are routinely embedded in an OWL ontology. It is this ability to define rich relationships between classes that gives an OWL-encoded ontology its power.

We see other problems that render data level integration in the STEP or IFC mould less effective than might otherwise be the case. To understand this position it is necessary to consider the way in which data integration mandates a considerable degree of work up-front. This is required in order to agree upon standards, construct a schema for integration and adapt applications to the standards etc., all before any benefits are realised. These issues become more onerous the larger the scope of agreement one is trying to achieve (inter-organisational, national, international etc.). Finally, for large international standards efforts, agility is something of a problem. Once the standard is agreed, changing it can take a considerable amount of time, which in an age of rapidly evolving business needs can turn a formerly helpful system into a hindrance (Boddy *et al.*, 2007).

The use of an ontology or multiple ontologies of the construction domain could act as a semantic abstraction layer above current standards and models to further integrate project data in a more intelligent fashion. For example, taking the point on views from Table 7.2, we believe an ontology with mappings into the underlying data models could be used to provide a more intuitive view of project data for any given actor based on their particular disciplinary concepts and terminology. That same ontology could also provide the view for an actor from a different discipline, based on the relationships explicated within the domain ontologies and between them and the core or upper ontology providing links to the appropriate terminology for the same data items. This type of 'translation' function becomes more compelling when used to view initial project briefs or client constraints and later when viewing the rationale for changes as it helps all actors to understand the reasoning involved in a language they can comprehend easily. Indeed extensions to the IFCs have been proposed to map them into ontologies for the construction domain to improve the semantic interoperability of BIM models in just such a way (Yang and Zhang, 2006). The mapping of ontology concepts into the current data model specifications would be performed initially in a semi-automated fashion using existing tools (Amor, 2004) for mapping between the data standards themselves, suitably modified for the task.

Taking the schema evolution point from Table 7.2, we would argue for a controlled process for the update and revision of both core and domain ontologies (Rezgui, 2007d). This process of continuous refinement of the ontologies means that at domain level and above, they should evolve naturally with use by the actors in the domains. However, ontologies would not necessarily help with the evolution of the design-level information schema (i.e. the drawings or models created by designers), a problem which would remain to be resolved by further research. By contrast, neither STEP nor the IFCs address these issues directly and the relatively static release-based versions of the standards limit what might otherwise be a route to domain-level evolution to a crawl.

With respect to issues around views over model data, the Express suite of languages employed by the STEP standards include Express-X, which can be used to create so called 'views' of Express-based STEP models. These views are essentially new models based on a new schema which has had the elements redundant for the current task or context removed or otherwise transformed by aggregation, decomposition etc. Express-X is used in this scenario to define the way in which the base or input schema[s] (and models based on it) is/are related to the view schema. Language constructs allow for the definition of rules about how to derive objects and properties in the view schema from the input schema[s]. In the authors' opinion, Express-X adds nothing in terms of the internal semantic expressiveness of either the base or view schemas and a dedicated mapping has to be written for each set of base and view schemas.

Where unstructured project information is concerned, the use of ontologies in tandem with other techniques drawn from information retrieval/extraction could be used to automatically infer links between the structured and unstructured information and indeed between items of unstructured information, based on the links defined in the ontology. These links lend a greater degree of context to each item relative to the project as a whole. Benefits may also be derived from uncovering previously unseen linkages between various elements of project data using such analysis methods. The eCognos (Lima *et al.*, 2005; Rezgui, 2006) project for example, developed and used a construction oriented ontology to augment the services that it offered as part of the collaborative knowledge management environment also developed on the project. The FUNSIEC project reviewed numerous European semantic resources, compiling them into an educational 'Experience Centre' and further conducting a feasibility study into the production of what the project termed an 'Open Semantic Infrastructure for the European Construction Sector' (OSIECS) (Barresi *et al.*, 2005).

7.5 Reasons behind the low adoption of IFCs

A number of studies have been reported in the literature describing various theories and models related to information technology adoption, diffusion and innovation into the workplace and across industry. Some of these theories describe transition processes and mechanisms, including Rogers's stage model of Diffusion of Innovations (DoI) in organisations (Rogers, 1995); whereas others define causality among factors to predict successful transition of a technology, including Davis's Technology Adoption Model (TAM), (Davis, 1993). TAM argues that end-user acceptance and use of information systems innovations is influenced by their beliefs regarding the technology. In particular, it proposes that perceived usefulness and perceived ease of use influence the use of information systems innovations and that this effect is mediated through behavioural intentions to use (Davis, 1989). The model highlights the critical role of extrinsic motivation and, in particular, expectations of task-related performance gains in end-users' adoption and use of information systems innovations (Davis, 1989). Roger's Diffusion of Innovation Theory argues that the rate of adoption of a technology is influenced by a number of factors, including: relative advantage (the degree to which potential adopters see an advantage for adopting the innovation), compatibility (the degree to which the innovation fits in with potential adopters' current practices and values), complexity (the degree of the innovation's ease of use), trialability (the degree to which potential adopters have the availability of 'testing' before adopting), and observability (the degree to which potential adopters are able to see observable results of an innovation).

Considering both models (TAM and DoI) in the context of IFCs, the authors discuss the various factors that may provide an initial explanation to the low adoption of IFCs:

- *Perceived usefulness of IFCs:* the research community and CAD software industry has failed in convincing the user community about the usefulness in adopting IFCs. This can be attributed to: (a) ineffective campaigning and awareness raising from the CAD vendors who tend to be driven by market share and competitiveness concerns; (b) the gap between the construction IT research community and the construction end-users; (c) the nature of the industry which is dominated by a large proportion of SMEs that operate in a survival mode and do not have the resources to consider, investigate or invest in sophisticated and costly solutions.
- *Ease of use of the IFCs (also related to Rogers's complexity factor):* the user community has in its majority used CAD to mimic traditional and manual ways of producing drawings. As such, ease of use is intrinsically linked with the complexity associated with the adoption of an object-oriented approach to the production of project documentation. It requires a paradigm shift whereby users have to adopt a lifecycle approach to data and information integration and shift from document (drawing) production to maintaining and enriching a building information model that serves as a basis to generate various consistent documentation, including drawings. The general feeling of the industry is that the adoption of the IFCs would require a steep learning curve.
- *Relative advantage:* it can be argued that, based on the above facts, the research and CAD vendor community (as they are ultimately the ones who are in charge of implementing the IFCs) have failed in convincing the user community about the advantages resulting from the adoption of the IFCs. In fact, there are limited semantic CAD solutions, and these have been mainly used in research and academic circles. The user community does not therefore see a relative advantage in adopting the IFCs. Indeed, there is a belief that the technology is not yet mature and reliable and that there is an inherent adoption risk.

- *Compatibility*: this is also a major concern as the IFCs require a paradigm shift and a migration from document centred to building information model-oriented approaches to design. Real incompatibility concerns have been raised, as there is a lack of preparedness and availability of a full and complete suite of IFC compatible ICT solutions to support the complex design stages.
- *Trialability*: whilst there are many commercial applications that can import or export IFC format files, their native formats are entirely different and proprietary. Thus trailability is difficult as existing demonstrators are mainly academic prototypes that can hardly be put between the hands of practitioners as these: (a) are not stable and robust enough; (b) have mainly been developed with open source or non-industry friendly platforms/solutions (e.g. object-oriented database systems); (c) require an education/training programme prior to testing so that the users would grasp the underlying concepts.
- *Observability*: A number of efforts have been made in this respect. For instance, the ATLAS project team in the early 1990s have released an interesting video that illustrates the advantages of adopting product data technology. However, this did not reach the whole user community and should have been articulated around a training and education initiative aimed at potential users from large, medium and small construction firms. Instead, this was mainly released and used amongst the research and academic community.

Addressing the above factors should facilitate (a) quicker uptake of a technology, before attention from adopters fades away and (b) critical mass of adoption, as there must be sufficient users and sufficient software providers that support a standard/technology before it becomes economically viable. Hence, 'timing' and 'critical mass' emerge as key success factors.

7.6 Requirements for a successful ontology

The authors would argue that the development of an ontology should factor in the above considerations and, hence, be embedded in a wider initiative aimed at raising awareness of the construction user community through education, training, and demonstrations.

Also, given the following factors: (a) the fragmented and discipline-oriented nature of the construction sector; (b) the various interpretations that exist of common concepts by different communities of practice (disciplines); (c) the plethora of semantic resources that exist within each discipline (none of which have reached a consensual agreement); (d) the lifecycle dimension of a construction project with information being produced and updated at different stages of the design and build process with a strong information-sharing requirement across organisations and lifecycle stages; a suitable ontology development methodology should accommodate the fact that the ontology should be specific enough to be accepted by practitioners within their own discipline, while providing a generic dimension that would promote communication and knowledge sharing amongst these communities.

A critical analysis of the semantic resources available in construction, ranging from taxonomies to thesauri, combined with an understanding of the characteristics of the sector, have helped formulate a set of requirements that ought to be addressed in order to maximise the chances of a wide adoption of any ontology project in the construction sector. These requirements are listed below (Rezgui, 2007d):

1   The ontology should not be developed from scratch but should make as much use as possible of established and recognised semantic resources in the domain.
2   The ontology should be built collaboratively in a multi-user environment: the construction sector involves several disciplines and communities of practice that use their own jargon and have specialised information needs.
3   There is a need to ensure total lifecycle support, as the information produced by one actor within one discipline should be able to be used by others working in related disciplines.
4   The ontology must be developed incrementally involving the end-users. This is important given the multi-disciplinary and multi-project nature of the industry, and the fact that each project is a one-off prototype.
5   The ontology should be flexible and comprehensive enough to accommodate different business scenarios used across projects and disciplines.
6   The ontology should be user-friendly, i.e., easy to use and providing a conceptualisation of the discipline/domain being represented that embeds the technical jargon used in the sector.
7   The ontology should be a living system and should allow for future expansion.

Also, while domain conceptualisations such as IFC had a strong application in CAD, ontologies should take a more broader perspective as they can be used as an effective mean to share and manage knowledge.

7.7 Ontology development methodologies

As reported by (Corcho *et al.*, 2003) and (Cristani and Cuel, 2005) a variety of methodologies have been developed for dealing with ontologies. These include:

- Methodologies for ontology building (Blazquez *et al.*, 1998; Gruninger and Fox, 1995; Holsapple and Joschi, 2002);
- Methodologies for ontology reengineering (Klein, 2001);
- Methodologies for ontology learning (Cimiano and Volker, 2005; Kietz *et al.*, 2000);

- Methodologies for ontology evaluation (Gomez-Perez, 2001; Guarino and Welty, 2000; Guarino *et al.*, 1999; Kietz *et al.*, 2000)
- Methodologies for ontology evolution (Klein, 2001; Klein *et al.*, 2002; Noy and Klein, 2002); and,
- Methodologies for ontology merging (Klein, 2001; Stumme and Maedche, 2001).

The co-existence of numerous methodologies suggests that a consensual methodology is difficult to establish due to the difficulty of developing a methodology adaptable to different applications, sectors and settings (Rezgui, 2007d). For instance, most of these methods and methodologies do not consider the collaborative and distributed construction of ontologies. Few methods (Holsapple and Joschi, 2002) include a proposal for collaborative construction of consensual ontologies primarily using domain experts. They include a protocol for agreeing new pieces of knowledge with the rest of the knowledge architecture.

Existing methodologies for building ontologies can be distinguished by two main aspects:

(a) the degree of dependency of the ontology on its application field (i.e., application dependent, semi-application dependent or generic); and/or
(b) the mechanism for deriving the ontology: generalisation or specialisation.

Furthermore, in relation to aspect (a), some methods use the application field as a starting point for building the ontology and are therefore application-dependent. This is the case of the KACTUS project (Bernaras *et al.*, 1996) and the On-To-Knowledge methodology (Staab *et al.*, 2001). Other methodologies can be described as semi application-dependent as they have the aspiration to be generic while using a given application as a reference. This is the case of the Gruninger and Fox methodology (Gruninger and Fox, 1995) and the method based on Sensus (Swartout *et al.*, 1997). Finally, some methodologies are generic, hence application-independent, since the ontology development process is totally independent of the uses of the ontology. This is the case of Uschold and King methods (Uschold and King, 1995), and Methontology (Fernandez-Lopez *et al.*, 1999). With few exceptions, Methontology (Fernandez-Lopez *et al.*, 1999) and Tove (Gruninger and Fox, 1995), it is worth noting that most reported methodologies proceed by building the ontology as a one-off exercise (Holsapple and Joshi, 2002).

In terms of the mechanism to derive the ontology (aspect b), if we compare the KACTUS and the Sensus methods, in the former the ontology is built by means of an abstraction process from an initial knowledge base, while, in the latter, an ontology skeleton (presented in the form of a simple taxonomic structure) is automatically generated from a huge ontology developed by merging and extracting information from existing electronic resources, including the semantic database of WordNet and an English dictionary.

A plethora of ontology management tools and environments have also been developed providing, essentially, graphical interfaces for managing ontologies augmented with various functionalities (e.g., ontology integration and merging) or reasoning capabilities. The latter category includes KAON2 (http://kaon2. semanticweb.org/), SWOOP (Kalyanpur *et al.*, 2005), OntoEdit (Sure *et al.*, 2002), Oiled (Bechhofer *et al.*, 2001), Protégé (Kietz *et al.*, 2000; Noy *et al.*, 2000) and Ontolingua (Farquhar *et al.*, 1996). Furthermore, as described in (Guarino and Welty, 2000) the former category (environments supporting ontology integration and merging) includes FCA-Merge (Stumme and Maedche, 2001), OntoMorph (Euzenat, 1996), Prompt (Noy and Musen, 2000), and Chimaera (McGuiness *et al.*, 2000).

These ontology management environments usually provide support for one or several ontology representation and structuring models, including OIL (Horrocks and van Harmelen, 2001; Partridge, 2002), RDF (Resource Description Framework), and OWL (Web Ontology Language – based on RDF).

7.8. Conclusion

The chapter argues that ontologies provide a richer conceptualisation of a complex domain such as construction compared to existing product data standards. The reasons behind the low adoption of IFCs are discussed followed by requirements for a successful construction ontology development and large-scale adoption. Finally, the chapter discusses existing ontology development approaches and techniques.

The following chapter reports on a methodology used to develop a construction domain ontology, taking into account the wealth of existing semantic resources in the sector ranging from dictionaries to thesauri. A construction industry standard taxonomy (IFCs) was used to provide the seeds of the ontology, enriched and expanded with additional concepts extracted from large discipline-oriented document bases using information retrieval techniques.

8 Construction ontology development

Ontology development approach; Ontology architecture definition; Semantic resources selection; Ontology modules construction; Testing and validation of the ontology; Conclusion.

8.1 Ontology development approach

The choice of any ontology development methodology is very much dependent on (a) the nature and characteristics of the targeted domain and its applications (i.e. various disciplines and related business processes), (b) the resources and development time available, and (c) the required depth of analysis of the ontology (Rezgui, 2007d). A critical analysis of the semantic resources available for the construction industry, including those listed in Table 7.1, combined with an acute understanding of the characteristics of the construction sector, have helped formulate a set of requirements that ought to be addressed in order to maximise the adoption and diffusion chances of any ontology project (see section 7.6).

Moreover, given the following factors discussed earlier in the book:

(a) the fragmented and discipline-oriented nature of the construction sector;
(b) the various interpretations that exist of common concepts by different communities of practice (disciplines);
(c) the plethora of semantic resources that exist within each discipline (none of which have reached a consensual agreement);
(d) the lifecycle dimension of a construction project with information being produced and updated at different stages of the design and build process with a strong information sharing requirement across organisations and lifecycle stages,

a suitable methodology should accommodate the fact that the ontology should be specific enough to be accepted by practitioners within their own discipline, while providing a generic dimension that would promote communication and knowledge sharing amongst these communities.

The ontology should be developed incrementally, in a collaborative way, involving representatives from the various disciplines, in order to promote ontological commitments and provide a mechanism whereby stakeholders share and exchange their perspectives and expertise (Hahn and Schulz, 2003). As mentioned earlier, textual documents have an important role in sharing and conveying knowledge and understanding. The methodology, illustrated in Figure 8.1, comprises the following stages: domain scoping, ontology architecture definition, candidate semantic resources selection, ontology modules development, ontology testing and validation, and ontology maintenance. For pragmatic reasons, these stages were grouped into four main phases and described as such in the following sections: Phase 1 (Domain scoping and architecture definition), Phase 2 (Candidate semantic resources identification), Phase 3 (Ontology modules construction), and Phase 4 (Ontology validation and maintenance).

Phase 1 addresses requirements 2, 3 and 7 as formulated in section 7.6. It helps identify the boundary conditions of the construction domain, and defines a suitable ontology development architecture. Phase 2 addresses requirement 1. It takes as input the ontology architecture from Phase 1, as well as the domain boundary conditions, and identifies relevant candidate semantic resources. Phase 3 addresses requirements 2, 3, 5 and 6. It takes as an input the identified candidate semantic resources from Phase 2, as well as the ontology architecture from Phase 1, and produces the ontology. Finally, Phase 4 addresses requirements 6 and 7. It tests, validates and maintains the ontology resulting from Phase 3. The decision was taken that the ontology would conform to an underlying knowledge model involving concepts, attributes and relations, defined using OWL.

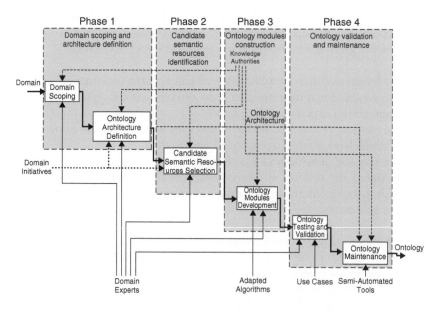

Figure 8.1 The various stages of the methodology (adapted from Rezgui, 2007d).

8.2 Ontology architecture definition

The scope of the ontology to be developed can be summarised by the following statement (Lima *et al.*, 2005): 'In the context of a Project, a group of Actors uses a set of Resources to produce a set of Products following certain Processes within a work environment (Related Disciplines) and under well defined conditions (Technical Topics).' As such, the scope of the ontology can be conveyed through seven major themes: (1) project, (2) actor, (3) resource, (4) product, (5) process, (6) technical topics (conditions), and (7) related disciplines (work environment).

The first five domains coincide with major themes in the IFC model. The last two domains include related issues that are not directly covered by the IFC.

Given the requirements formulated in section 7.6, the eCognos ontology is structured into a set of discrete, core- and discipline-oriented sub-ontologies referred to as modules (Figure 8.2). Each module features a high cohesion between its internal concepts while ensuring a high degree of interoperability between them. These are organised into a layered architecture with, at a high level of abstraction, the core ontology that holds a common conceptualisation of the whole construction domain enabled by a set of inter-related generic core concepts forming the seeds of the ontology. These generic concepts enable interoperability between specialised discipline-oriented modules defined at a lower level of abstraction. This middle layer of the architecture provides discipline-oriented conceptualisations of the construction domain. Concepts from these sub-ontologies are linked with the core concepts by generalisation/spe-

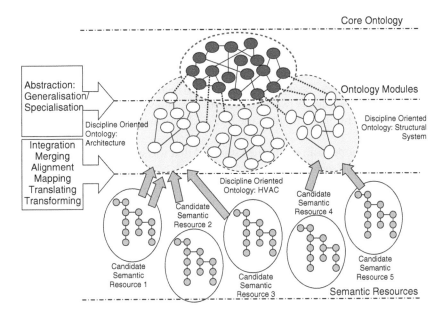

Figure 8.2 The eCognos ontology architecture (adapted from Rezgui, 2007d).

cialisation (commonly known as IS-A) relationships. The third and lowest level of the architecture (Figure 2) represents all semantic resources currently available, which constitute potential candidates for inclusion into eCognos either at the core or discipline level.

8.3 Semantic resources selection

There are a large variety of available semantic resources that can form the basis for building the eCognos core ontology (see section 7.1). These range from classification systems to taxonomies. The latter do deserve particular attention as argued in (Guarino and Welty, 2000; Welty and Guarino, 2001). One of the principal roles of taxonomies is to facilitate human understanding, impart structure on an ontology and promote tenable integration. Furthermore, properly structured taxonomies:

- help bring substantial order to elements of a model;
- are particularly useful in presenting limited views of a model for human interpretation; and,
- play a critical role in re-use and integration tasks. Improperly structured taxonomies have the opposite effect, making models confusing and difficult to re-use or reintegrate (Welty and Guarino, 2001).

IFC, being more recent and also the closest taxonomy currently in use in the sector, is therefore the preferred candidate semantic resource that can provide the skeleton on which such a core ontology can be built. The 85 major concepts composing the IFC kernel (as described below) therefore formed the basis upon which to build the skeleton of the eCognos ontology. The wealth of information in related semantic resources, including BS 6100, is used to enrich the eCognos ontology. As argued in the previous section, only generic concepts used across disciplines should be included in the core ontology. These concepts should hold minimal semantics necessary to establish a sound foundation for the ontology, supporting interoperability between the various discipline-oriented sub-ontologies.

The main advantages in using the IFC model include the following:

- the IFC model is an established standard that was strongly promoted in the construction sector;
- each IFC entity has a defined set of mandatory attributes that define its minimal structure;
- a taxonomy as argued in (Guarino and Welty, 2000; Welty and Guarino, 2001) provides an ideal backbone for any ontology project.

On the other hand, the IFC entities are regarded as 'very abstract concepts' that do not reflect the terms and language (jargon) used in end-users' daily work. In other words, they are not semantically rich enough to provide a true and user-friendly conceptualisation of their discipline, and are therefore not sufficient to form a suitable ontology.

As far as the discipline-oriented ontologies are concerned, if a related concep-tualisation of the discipline exists, including a taxonomy, then this is selected as a candidate for the ontology. This process is detailed in the following section.

8.4 Ontology modules construction

This stage involves the technical evaluation of the previously identified candidate semantic resources by domain experts, with the objective of determining their suitability for re-use as discipline ontologies in accordance with the proposed modular architecture. In theory, once the semantic resources have been selected, several operations can be used in constructing an ontology. These include:

- *Integration:* The integration operation involves the integration of one ontology into another through their overlapping parts. While the inte-gration operation provides a linkage and tight coupling between the two ontologies, their initial scope is preserved.
- *Merging:* The merging operation involves creating a new ontology from two or more existing ontologies, and bringing them as a result into mutual agreement, making them consistent and coherent. The initial ontologies cease existing in isolation and are replaced with the new ontology.
- *Translation:* This operation involves changing the representation formalism of an ontology while preserving its semantics.
- *Transformation:* This operation involves changing the semantics of an ontology to make it suitable for purposes other than the one for which it was originally conceived.

While these operations can be used to address the co-existence of several candi-date semantic resources to construct and converge into a unique domain ontol-ogy (as illustrated in the lower layer of the architecture in Figure 8.2), a tax-onomy should be used wherever available as the preferred semantic resource for building a discipline-oriented ontology (Guarino and Welty, 2000; Welty and Guarino, 2001). In the absence of a taxonomy, the semantic resource that has clear support from industry and discipline practitioners should be retained. This can then be enriched and expanded using additional terms and relationships.

A particular approach was adopted for expanding the ontology. This involves selecting and making use of a large documentary corpus used in the discipline and ideally produced by the end-users. Expanding or building an ontology from index terms extracted from commonly used documents in a given discipline requires a few operations organised into steps. These are described below and are applied to each document of the selected documentary corpus.

Step 1: document cleansing

This step aims at reducing the document to a textual description that con-tains nouns and associations between nouns that carry most of the document semantics. This involves the following steps:

- Lexical analysis of the text in order to deal with digits, hyphens, punctuation marks, and the case of letters;
- Elimination of stopwords: this has the objective of filtering out non-content words, such as 'the' and 'of'.

Step 2: keyword extraction

This step aims at providing a logical view of a document through summarisation via a set of semantically relevant keywords. These are referred to, in this stage, as index terms. The purpose is to gradually move from a full-text representation of the document to a higher-level representation. In order to reduce the complexity of the text, as well as the resulting computational costs, the index terms to be retained are all the nouns from the cleansed text.

Several authors (Baeza-Yates and Ribeiro-Neto, 1999; Broglio *et al.*, 1995) argue that nouns, as opposed to verbs, adjectives and adverbs, carry out most of the semantics of a text document. This is assumed to be the case for the construction domain where technical documentation refers to noun-based terms, such as products, construction materials and resources. It is argued in this research that the type of semantics carried out by verbs in the construction sector describes and refers to relationships between ontology concepts. These relationships are identified in Step 4 of the proposed approach.

However, concepts in the construction sector are sometime carried out by more than one term, such as 'curtain wall' and can therefore be conveyed by bigrams. More generally, the notion of an n-gram can be found in the literature when a sequence of n terms is regarded as a single object. While n-grams play a key role in capturing term collocations (Tomovic *et al.*, 2006), it is also worth noting that computation of n-gram probabilities does not accurately cater for rare term collocation events (Omelayenko, 2001).

The approach presented here combines nouns co-occurring with a null syntactic distance (i.e. the number of words between the two nouns is null) into a single indexing component, regardless of their collocation frequency. These are referred to as noun groups (non-elementary index terms).

The result of this step is a set of elementary and non-elementary keywords that are representative of the discipline being conceptualised.

Step 3: integrating index terms into core and sub-ontology

Two types of concept integration are possible (Omelayenko, 2001): integration at both conceptual and syntactic levels. Concept level integration requires inference over the domain ontology to make a decision about integration of a particular pair of concepts. Syntactical integration defines the rules in terms of class and attribute names to be integrated. Such integration rules are conceptually blind but are easy to implement and develop (Omelayenko, 2001). The approach used for concept level integration makes use of a pivotal semantic resource, ideally a thesaurus or a taxonomy as illustrated in Figure 8.3.

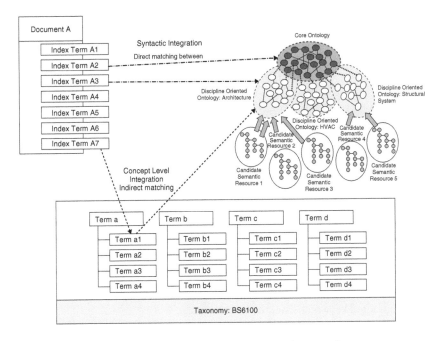

Figure 8.3 Method for concept integration making use of a pivotal semantic resource (adapted from Rezgui, 2007d).

As highlighted in (Baeza-Yates and Ribeiro-Neto, 1999) a glossary or thesaurus can provide a controlled vocabulary for the extension of the ontology based on the identified keywords. A controlled vocabulary presents the advantage of normalised terms, the reduction of noise, and the possibility of turning keywords into concepts with clear semantic meaning. The construction BS6100 glossary, which is also structured as a taxonomy, was used.

For each identified keyword, it is important to quantify the degree of importance (in terms of semantics) it has over not only the document but also the entire documentary corpus selected for the given discipline. The following formula, known as 'Term frequency-inverse document frequency' (tf-idf) (Baeza-Yates and Ribeiro-Neto, 1999; Salton and Buckley, 1988), is used:

$$W_{i,j} = f_{i,j} \times idf_i \qquad (8.1)$$

Where $W_{i,j}$ represents the quantified weight that a term t_i has over the document d_j; $f_{i,j}$ represents the normalised occurrence of a term t_i in a document d_j, and is calculated using Equation (8.2):

$$f_{i,j} = \frac{freq_{i,j}}{\max_{\text{for all terms in document}} freq_{term,j}} \qquad (8.2)$$

Where $freq_{i,j}$ represents the number of times the term t_i is mentioned in document d_j; maxfor all terms in document $freq_{term,j}$ computes the maximum over all terms which are mentioned in the text of document d_j; idf_i represents the inverse of the frequency of a term t_i among the documents in the entire knowledge base, and is expressed as shown in Equation (8.3):

$$idf_i = \log \frac{N}{n_i} \qquad\qquad (8.3)$$

Where N is the total number of documents in the knowledge base, and ni the number of documents in which the term t_i appears. The intuition behind the measure of $W_{i,j}$ is motivated by the fact that the best terms for inclusion in the ontology are those featured in certain individual documents, capable of distinguishing them from the remainder of the collection. This implies that the best terms should have high term frequencies but low overall collection frequencies. The term importance is therefore obtained by using the product of the term frequency and the inverse document frequency (Salton and Bukley, 1988).

The following hypotheses apply when comparing concepts from the ontology with extracted keywords from the documents:

• If an ontological concept and a document keyword have the same name or a common stem then they can be considered as semantically equivalent. This is motivated by the nature of the construction jargon where variations of the same noun are used in various contextual situations. This is best illustrated by 'architecture' where variations of the concept exist, including 'architectural' used to express an 'architectural point of view' or an 'architectural solution'. While when looked at in isolation the extracted keyword 'architectural' will be treated as semantically equivalent to the ontology concept 'architecture', the non-elementary index term 'architectural solution' (as explained earlier) will in fact be considered as a potential new concept for the ontology, to be validated by knowledge experts.

• If an ontological concept and a document keyword have distinct names, both concepts are looked up in the BS 6100 glossary. The algorithm checks first if both concepts are related through a specialisation/generalisation mechanism. In the affirmative, the keyword is added as such to the ontology in accordance with the identified relationship, and is marked as 'new', which means that it will only be made officially part of the ontology once approved by the relevant knowledge expert(s). If the ontological concept and the document keyword are not related through a specialisation/generalisation relationship, then the quantified semantic weight of the index term (Equation 8.1) is used as a criterion to decide whether it should be added as a concept to the ontology. At this stage, only index terms with a relatively high semantic weight were retained and added to the ontology with no relationship to other concepts defined at this stage.

In order to devise a threshold for the term weighting factor, an experiment was conducted on the 'Architecture domain' ontology, whereby domain experts were asked to analyse the retrieved concepts. A collection of 56 documents was used, out of which 480 concepts were extracted and ordered by the decreasing value of their highest associated term weighting factor (computed over each document of the related document collection). Out of the 480 extracted concepts, 447 concepts were new and not included in the initial existing taxonomy in Architecture. The three selected architect domain experts for the experiment analyzed and retained the top 468 concepts arguing that the remaining 17 concepts were not semantically relevant. This revealed that below the value of 0.22 of the term weighting factor, related concepts were either semantically equivalent to the ones defined above this value, or were deemed not worth including in the ontology. This figure (0.22) was then used as a threshold for retaining the concepts of the three other domain ontologies.

Finally, the properties of the new ontological concepts are either refined or defined for new concepts that have been integrated into the ontology (each property defined in the concept will receive, if applicable, ranges, restrictions, etc.). This task in undertaken manually by the knowledge experts.

Step 4: ontology concept relationship building

The next step in the methodology includes building the relationships connecting the concepts, including those that have not been retained in the previous stage. Concept relationships can be induced by patterns of co-occurrence within documents. As described in (Baeza-Yates and Ribeiro-Neto, 1999), overall relationships are usually of a hierarchical nature and most often involve associations between 'broader' and 'narrower' related terms. Broader term and narrower term relationships define a specialisation hierarchy where the broader term is associated with a class and its related narrower terms are associated with specialised instances of the class. While broader and narrower terms relationships can be defined automatically, dealing with related term relationships is much harder. One reason for this is that these relationships depend on the specific context and the particular needs of the group of users, and are thus difficult to establish without the knowledge provided by specialists. We therefore distinguish three main types of relationships:

- Generalisation/specialisation relationship (e.g., Wall can be specialised into separation wall, structural wall, loadbearing separation wall).
- Composition/aggregation relationship (e.g., Door is an aggregation of a frame, a handle, etc).
- Semantic relationship between concepts (e.g., a beam supports a slab, and a beam is supported by a column).

The last two categories above are addressed in this step. The process is semi-automated in that relations are first identified automatically. Contributions from knowledge specialists are then requested to qualify and define the identified relations. In order to assess the relevance of relationships between concepts, an approach that factors the number of co-occurrences of concepts with their proximity in the text is adopted. This is known as the 'Metric Clusters' method (Baeza-Yates and Ribeiro-Neto, 1999) (Equation 8.4). This proceeds by factoring the distance between two terms in the computation of their correlation factor. The assumption being that terms which occur in the same sentence, seem more correlated than terms that appear far away.

$$C_{u,v} = \sum_{t_i \in V(S_u)} \sum_{t_j \in V(S_v)} \frac{1}{r(t_i, t_j)} \tag{8.4}$$

The distance $r(t_i, t_j)$ between two keywords t_i and t_j is given by the number of words between them in the same document. $V(S_u)$ and $V(S_v)$ represent the sets of keywords which have S_u and S_v as their respective stems. In order to simplify the correlation factor given in Equation 8.4, it was decided not to take into account the different syntactic variations of concepts within the text, and instead use Equation 8.5, where $r(t_u, t_v)$ represents the minimum distance (in terms of the number of separating words) between concepts t_u and t_v in any single document

$$C_{u,v} = \frac{1}{Min[r(t_u, t_v)]} \tag{8.5}$$

The following exception should be noted: in the case where the minimum distance between two terms is null the correlation factor will return '∞'. When this exception arises, the correlated terms are considered as candidates to form a composite term. The assumption taken in the computation of Equation 8.4 is that the closer the correlation factor is to 1, the stronger the correlation between the two terms is likely to be. Based on the correlation factors, index terms are linked to relevant concepts from the ontology with a blind relationship. In order to establish a threshold for the correlation factor, it was decided that only terms co-occurring within the same sentence should be considered. An analysis of a sample set of documents showed the average size of a sentence to be 12 words. $C_{u,v}$ has therefore been given a threshold of 0.1 (maximum distance between two terms co-occurring in a 12-word sentence). The decision to retain the average size of a sentence was motivated by the results of the experiment as the threshold resulting from the use of a '12-word sentence' helped identify a limited but most of the time justified number of relationships.

The domain knowledge experts have the responsibility of validating the newly integrated index terms as well as their given names, and then defining all the concept associations that do not belong to the generalisation/specialisation category. First, these relationships were established at a high level within the core ontology, and then subsequent efforts have established relationships at lower levels within the discipline ontologies.

The use of discipline documents to identify ontological concepts and relationships was revealed to be the right strategy to construct the discipline sub-ontologies (Rezgui, 2007d). As far as relationships are concerned, the approach helped identify:

- relations that exist between concepts found in the initial taxonomies (which only included relations of type 'IS-A'). For instance IfcColumn, IfcBeam, and IfcSlab are concepts that existed in the initial taxonomy as a specialisation of IfcBuildingElement. Relationships between these concepts have now been enriched in the Architecture sub-ontology by the following new semantic links: Beam supports Slab; Column supports Beam.
- new relations brought by the inclusion of a substantial number of new concepts. For instance the new concept Opening has generated a number of new relations such as Opening *is performed on* Wall; Window or Door *is fitted on* Opening.

While the initial core ontology included 71 concepts, it has now been expanded and enriched in the light of the newly added concepts in the various sub-ontologies. The core ontology includes 114 concepts in its current version. A comprehensive version of the core ontology is illustrated in Figure 8.4. Also, it is

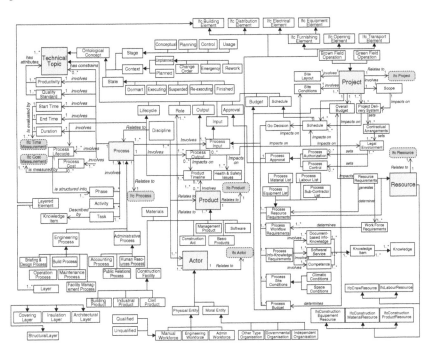

Figure 8.4 Snapshot of the eCognos core ontology (adapted from Rezgui, 2007d).

worth noting that while the concepts in the initial taxonomies identified the core product components within each discipline, the use of documents helped identify concepts that provide a level of completeness, which makes the sub-ontologies self sufficient to conduct end-users' business processes.

8.5 Testing and validation of the ontology

Validation of the ontology involved four organisations that each selected a knowledge intensive business unit in order to test and validate the developed ontology. The ontology was used through a set of knowledge management dedicated services delivered as part of the eCognos platform (Wetherill *et al.*, 2003). The testing of the ontology aimed at:

(a) establishing the completeness of the ontology in terms of conceptualising the targeted disciplines;
(b) assessing the relevance of the concepts and relationships; and,
(c) evaluating the usefulness and advantages afforded by the use of an ontology for knowledge management purposes, as opposed to traditional approaches based on keywords.

The main services of the eCognos platform that were used to validate the ontology include: Knowledge extraction, Knowledge indexing, and Knowledge searching services. The eCognos platform generates a Knowledge representation (KR) for every single Knowledge item (KI) managed by the system. The latter includes any text-based electronic object that holds construction-related semantics. When adding a new KI in the knowledge base, the knowledge extractor service is responsible for extracting the best relevant keywords from the KI and calculating their statistical weights. This set of keywords/weights is referred to as a semantic vector and is submitted to the ontology service in order to get the best ontological representation of the KI. An example of a semantic vector is illustrated hereafter:

{(airlock,0), (balcony,0), (Beam,0.05), (covering,0.1), (duct,0.5), (hinge,0.02), (lift well, 0.2), (shaft,0.3), (Tendons,0), . . .}.

The dimension of the vector is determined by the number of concepts in the ontology.

A KR is then built for the KI and stored in the eCognos repository. It is worth highlighting that this process is independent from the one illustrated in Figure 8.1. While the latter is used to develop the ontology, the former illustrates examples of the use of the ontology to generate ontological knowledge representations of KIs used on a daily basis by construction practitioners. Also, when searching for knowledge, the system submits the user's query in order to get the best ontological representation of the resulting semantic vector. The search service was used in particular to enable users to perform searches across knowledge items, including documents, and to allow them to retrieve those knowledge items and their associated representations based on the concepts used in the ontology.

Two main metrics were used to evaluate the eCognos ontology:

(a) performance evaluation in terms of response time delivered by the use of the ontology; and,
(b) the retrieval performance in terms of relevance of the retrieved document set through the use of ontology.

Four sample document sets (representing the four selected case studies) provided by the project construction end-users were used. These were gathered from several recently completed projects. Fifteen queries per case study have been formulated by discipline experts involved in the research. These were based on their own information needs experiences on projects, traditionally performed manually or using ad-hoc search facilities.

The experts were asked to identify manually the document set matching each formulated query (within the context of their discipline). While the identification process was achieved in a matter of days, this did reflect (timescale wise) the current information and document search practices in the construction industry. These identified relevant document sets have then been used as a basis to compute the average recall (the fraction of the retrieved relevant documents) and precision measures (the fraction of the retrieved documents which is relevant). These have been quantified in two scenarios:

(a) through the simple use of index terms; and,
(b) through the use of the developed ontology.

In the first scenario, a full-text summarisation of the documents and queries was preformed, while in the second scenario the summarisation relied exclusively on concepts of the ontology. The discipline experts that generated the queries in the first place and identified the relevant document sets have been involved in the relevance assessment work.

In terms of response time, the difference between the two approaches (with and without the use of ontology) was comparable, and hence is not reported in the paper. The difference was in the retrieval performance provided by the ontology (Rezgui, 2007d). The description of the field trials and testing scenarios are beyond the scope of this chapter (related information can be found in (eCognos Consortium, 2003)). The main results of the evaluation are summarised below:

• The issue of the comprehensiveness of the ontology was raised, as a number of concepts were reported missing in some of the discipline ontologies. While acknowledging potential limitations of automated IR techniques, this suggests that either the selected term-weighting threshold was too high, or that the scope and coverage of the document corpus used to build the ontology supplied by the participating companies were not comprehensive enough. This can be addressed by lowering the term-weighting threshold in consultation with domain experts while enlarging the documentary corpus with further documents drawn from projects that exhibit distinct characteristics in terms of project type, construction system, choice of materials and geographical location.

- The assumption adopted earlier, whereby most construction semantics is expressed through elementary or non-elementary concepts composed exclusively of nouns, was challenged through the implementation process as we did encounter concepts that referred to verbs such as 'Reinforced Concrete'. While Stage 1 failed to detect such concepts, domain specialists, when asked to qualify the relationship between related elementary concepts, have ultimately identified these. In this particular example 'Reinforced Concrete' is created by objectifying the relationship between 'Concrete' and 'Steel Bar'. Based on the four targeted domain ontologies, the inclusion of verbs when combining words (nouns and verbs) co-occurring with a null syntactic distance into a single indexing component triples the number of retained concepts (Rezgui, 2007d). Based on the experience of this first iteration, only a small number of concepts involving verbs as elementary or non-elementary concepts (exactly 9) were reported missing. The inclusion of verbs would have put an additional burden on knowledge experts as these would have to analyse the 1496 additional concepts to retain only a few. It is therefore argued that while this can cause potential sources of concept loss, knowledge experts' time saving gained by restricting index terms to elementary or composite nouns justifies the choice of this option. It is also believed the latter stages of the approach, in particular Stage 4, can allow these to be identified and created.
- Despite reporting missing concepts, the search functionality with ontology outperformed the traditional full-text search approach as more relevant document subsets have been retrieved. The knowledge representation technique, based on ontology, provides a more accurate user and machine interpretable summarisation of documents, as illustrated by the field trial results with significant improvements in the precision and recall factors.
- The use of integrated services articulated around a common in-house ontology promotes the wide adoption of common standards and the sharing of a common understanding of terms and concepts. However, subtle variations of the semantics of certain terms may exist across companies. This suggests that discipline ontologies in the construction sector may need further refinement to be adapted to the norms and values of an organisation.
- Some business processes involve the use of concepts from more than one discipline ontology. While this is supported by the eCognos methodology through the generic core ontology, it triggers another issue in relation to supporting dynamic views or perspectives that involve concepts drawn from more than one discipline ontology.

The validation of the methodology stresses the importance of the incremental and iterative approach to building the ontology. A full reporting of the validation work can be found in Rezgui (2007d).

8.6 Conclusion

The chapter presented an ontology development methodology to support the information and knowledge needs of practitioners in the construction sector. A layered and modular approach was adopted to structure and develop the ontology. This was justified by the fragmented nature of the sector, organised into a variety of disciplines. The uniqueness of the methodology is illustrated by the combination of the following distinctive features:

- *The modular structure of the ontology:* The ontology takes into account the fragmented nature of the construction sector and its organisation into established disciplines. It, therefore, mirrors the discipline-oriented nature of the industry.
- *The support for the multiple interpretations of concepts across disciplines:* the construction disciplines have their own norms and values, reflected in the development of dedicated semantic resources. The proposed ontology architecture is modular, with sub-ontologies dedicated to each established discipline, federated by a core ontology defined at a higher level comprising generic concepts applicable across sectors and enabling reconciliation of different terminologies used across disciplines.
- *The collaborative nature of the ontology development process:* it supports parallel development of the various ontology modules while promoting their integration through the core ontology.
- *The iterative nature of the ontology development process:* given the large-scale dimension of the construction industry and its project-oriented nature, an iterative process that proceeds by refining and extending the ontology over time helps converge towards a complete and true conceptualisation of the domain.
- *The ontology development approach:* this is semi-automated and relies on discipline-oriented documentary corpuses to identify concepts and relationships using tf-idf and metric clusters techniques, which are then validated by human experts. It proceeds by re-using and building on existing semantic resources, improved and enriched with additional concepts drawn from discipline-oriented documentary corpuses.

The ontology developed to date is far from being complete, and will probably never be, as an ontology should be viewed as a living system. Similar efforts, such as the ISO STEP (STEP, 1994) project and its application to various industry sectors (including manufacturing), have taken almost a decade or longer to come to fruition. eCognos, or any other ontology development, is by no means different. The issue of the existence of a unique ontology for an entire sector remains open. This suggests that while the eCognos core ontology forms a robust basis for interoperability across the discipline-oriented ontologies, the latter will need adaptation and refining when deployed into an organisation and used on projects. Another issue that was raised is that related to the adoption of user-specific views or perspectives on the global ontology. In fact, in many instances, some actors might be required as part of their job to deal with more than one discipline ontology to conduct a task. This necessitates some flexible mechanisms that can enable the rapid combination of two or more discipline ontologies into a single view/perspective. This constitutes an interesting direction for future research.

9 Complex problem solving
The use of evolutionary algorithms

Why evolutionary algorithms; Examples of complex problems; The value of experience; A search space; The solution; Search algorithms; Genetic algorithms; Encoding the problem; The choice of encoding; Fitness assessment: the fitness function; Selection; Crossover; Mutation; Inversion; Convergence and results; Conclusion.

9.1 Why evolutionary algorithms

Many everyday problems have a huge number of feasible solutions. Often this can be as many as several billions, although almost inevitably, the people involved in making decisions about these problems are totally unaware of the complexity with which they are dealing. For such complex problems, the challenge for human decision makers is, ideally, to find 'the best' solution. If this is achieved it can be classed as optimisation, although this is a word which is used much too loosely, as we will discuss below. Very often, it is sufficient just to find a 'good' solution or a range of good solutions from which the decision maker(s) can choose.

Typically when faced with a complex decision-making problem, humans rely on heuristics to reduce the choice down to something more manageable. Usually these heuristics are based on experience but as we shall show, even for quite simple problems, it is not possible for the decision maker to gain enough experience to be able to develop heuristics which cover the complete search space.

In this chapter, we will start by briefly examining human decision making before moving on to look at how evolutionary algorithms can help to search through the mass of possible solutions. To do this the algorithms have to be very efficient. For example, if we take the a problem with 10^{55} possible solutions (Clarke *et al.*, 2009), if each solution takes a computer 1 second to evaluate, then an exhaustive search would take about 3^{47} years. As the sun is only expected to last for around 5 billion years, this is well beyond the life of the solar system. So to get a solution to such a problem within a sensible time span, say 1 hour, means that the algorithm has to achieve its search for good solution(s) by sampling just a minute fraction of the entire space of solutions.

9.2 Examples of complex problems

In this section, we will examine a few typical problems to show how something which appears to be simple can very quickly become highly complex. In the following text, the complete set of feasible solutions will be referred to as the search space.

9.2.1 Power generation

The first problem we will look at is one which we were asked to solve in the late 1990s for the UK government. At the time they were very concerned about the acidic emissions from fossil fuel-fired electricity-generating power plants and they wanted to develop strategies for mitigating the damage caused by these emissions. The results are confidential but the size of the search space can be discussed. At the time there were 40 major fossil fuel-fired electricity-generating stations, each of which typically could be operated in one of 80 states. We were required to produce a generating strategy which would be the best for every hour throughout a typical year, thus giving a search space of: $40 \times 80 \times 365 \times 24 = 28032000$.

Even with the slow computers that were available at the time, we were able to find a good solution within 20 minutes on a typical PC.

9.2.2 Built environment design

Design is an everyday activity which results in products which work and deliver value to their producers, purchasers and society but it is also hugely complex. Let us, for example, take a typical example of built environment design and limit it to just three participants, an architect, a structural engineer and a building services engineer. Typical areas of interest for these three participants are shown in Table 9.1.

If we assume that for each of the above there are 10 feasible choices (for some there will be more, for some less) and given that there are 18 decision variables then this gives a search space of 10^{18}. This is a very large search space and yet to do the calculation we have introduced some significant simplifications, for example no fire engineering, no contractor involvement etc., so in truth real design decision spaces are much larger. This is a theme which we will return to later in this chapter.

Table 9.1 Areas of interest example in building environment design

Architect Engineer	Structural Engineer	Building Services
Façade	Frame	Orientation
Foyer	Flooring systems	Cladding
Spaces	Stability	Thermal mass
Flow	Materials	Lighting
Usage	Foundations	Heating
Finishes		Acoustics
Aesthetics		

9.2.3 *Strategic decision making for fire and rescue*

We are currently working on a project for the Fire and Rescue Services of the UK (Clarke *et al.*, 2009). Rather than having a single fire service for the whole of the UK, the country is split into 58 areas each of which is covered by what is referred to as a brigade, this being the name for the local fire service. Each brigade then decides where to place its buildings to hold the fire engines and associated machines. So here immediately we have a decision, where to place the buildings and how many buildings should there be. Next comes the decision as to what sort of machines should be contained in each building, whether they should just be standard fire engines or whether there should be a mix of these plus special machines such as those used to rescue people from high buildings or rescue tenders to extricate people from road traffic accidents. The next decision is whether these machines should be manned full time or should some of them just be manned from, for example, dawn to dusk.

If we take the example of the South Wales brigade, this has 50 fire stations. If we assume they are considering 20 possible new sites this gives 70 possible locations. Assuming that the order is unimportant (i.e. having A, B and C open is the same as having C, B and A open) the total number of combinations is given by:

$$N_s = \frac{70!}{50!(7-50)} \approx 10^{17} \tag{9.1}$$

If we assume each station has 6 different vehicle and staffing options (this is a conservative estimate), then the number of possible combinations is 6^{50} for the 50 possible sites. As 6^{50} is approximately 10^{38} this gives the total number of combinations as ($10^{17} \times 10^{38} = 10^{55}$).

So this massive number is actually a conservative estimate.

The above examples serve to show that it is very easy for what at first sight looks like a simple problem to be, in reality, highly complex. In the next section we will look at how human beings apparently deal with such complex problems.

9.3 The value of experience

In the above section we have shown that relatively common problems can have a huge number of feasible solutions, far more than can be handled by human decision makers. This then leads to the question, if many decision making processes are so complex, how is it that human beings can cope? How do people manage to deal with complex problems and arrive at solutions that work?

The decision makers will tell you that they use their experience to sift through the myriad of possible solutions. They use what are called heuristics, based on experience, which are approximate methods sometimes called rules of thumb, to reduce the mass of possible solutions to something which is manageable. We can do some simple calculations to see how effective this approach is:

Let us assume there is a designer of buildings who works for forty years designing buildings. During these forty years, we will create a world in which there are no changes of technology or fashions so the building types stay static for the whole of the forty years. This designer is a fast worker so he designs, in collaboration with others, five buildings per year. So in his 40 years of experience, he will design 200 buildings.

To reach this figure we have had to make some significant simplifying assumptions and in reality the number would be less. In contrast, in the previous sections when we looked at the size of search spaces, we underestimated the size of the search space for a building by not including all possible factors and still arrived at a number of 10^{18} options. Not all of these will be feasible but even if only one per cent of them are this still leaves 10^{16}. As a fraction of this, 200 is 2^{12} per cent. So the concept of experience being a substitute for a full search is a myth.

So what happens in current design? The answer is that designers provide satisfying solutions, these being solutions that meet the constraints. So if they are buildings they resist the imposed loads and (mostly) keep the weather out but unless the design team is extremely lucky, they get nowhere near a 'good' design. This is especially true for designs which involve several disciplines. Each discipline will know something of the needs of the others but will tend to set its objectives and constraints in isolation. If these are ever overtly stated it will be at a design meeting but very often they are built into the discipline's design solution. Thus immediately there is a decision-making process in which there is a huge potential for the various disciplines to set their own sub-goals which, inadvertently restrict the other design teams and push them further away from 'good' solutions. This situation also applies to all complex, multi-discipline decision making processes.

9.4 A search space

The concept of a 'search space' is one that occurs throughout this chapter. A search space can be defined as the entire range of possible solutions to a problem. A potential search space is shown below in Figure 9.1. As can be seen the space has three dimensions. The two horizontal dimensions will typically represent the two variables that are present in the problem (for example, for an engineering problem these could be weight and cost) and the vertical dimension then represents the 'fitness' of the solution (i.e. how good the solution is). Note that the search space in this example is continuous (this is not always the case) and has many peaks of similar height. The problem therefore is to look around the search space and find the highest peak. For a simple two variable problem of the sort illustrated, this looks like an easy task, but most real search spaces are multi-dimensional and many have discrete variables and hence discontinuous search spaces which are a challenge for any algorithm.

9.5 The solution

To help people with decision, some form of support is required and inevitably given the typical complexity of problems, this has to be computer-based.

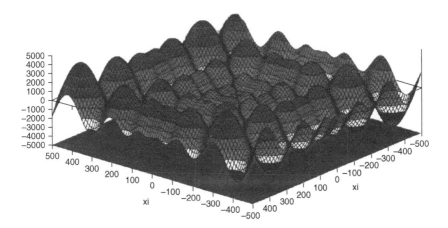

Figure 9.1 An example of a search space (adapted from Bradshaw, 1996).

So-called decision support systems are not new, having been around for almost as long as computers. Early decision support systems only allowed their users to evaluate one option at a time. While this is better than having no decision support, given the level of complexity that we have discussed above it is clearly not going to be much better than the decision makers using their heuristics. More recently, new forms of decision support system have been devised which include some form of search engine, the latter being capable of searching through and evaluating multiple options based on decision-making criteria which are typically pre-defined by the user. As alluded to above, these search engines have to be very efficient because they have to find good solutions while only sampling a tiny fraction of the total number of solutions. Some of the best examples of early development in this area are contained in the work of Parmee (1998, 2001) who mostly applied search techniques to the conceptual design of aircraft.

The fact that this form of decision support system is capable of undertaking a wide-ranging search introduces other useful features. Rather than just producing a single 'best' answer, such systems can be used to find multiple areas of interest within the search space, each of which contains several high performance solutions. So the user is then able to critically review the results that are produced and introduce their own expertise into the process to decide which of the solutions is the most promising (e.g. (Parmee, 2001, de Wilde *et al.*, 2009)). Additionally the user can then undertake a more localised search around the most promising solution. This may involve using simpler techniques such as hill-climbing. Alternatively, the user may choose to alter the decision-making criteria to examine the sensitivity of the results and particularly to examine the influence of any assumptions. This is easy to achieve and it also has the advantage that it helps the user to learn about the search space, thus giving him/her knowledge which could not be gained by any other means.

So such systems help users to learn more about the problem they are trying to solve and also guide them towards finding high-performance solutions. The execution times of such systems can vary substantially depending on the size and complexity of the replacement for the objective function, which is known as the fitness function.

9.6 Search algorithms

Evolutionary algorithms are a class of algorithms which have the following distinguishing features:

- Usage of a stochastic search process;
- An ability to find good solutions by only sampling a very small fraction of the search space;
- Employing a population of solutions rather than working on a single solution;
- Unlike traditional optimisation algorithms, they require relatively little information about the nature of the problem in order to be able to search effectively;
- A good ability for avoiding local optima;
- An ability to cope with constraints (typically by converting them into objectives);
- Coping with problems involving many objectives (some of which may conflict with one another).

Many of these features are coupled. For example the ability to generally avoid becoming stuck on a locally optimum solution is a function of the randomness introduced by stochastic search and the use of a population of solutions thus giving a wider coverage of the search space at any one time during the search process. Also, evolutionary algorithms cannot be guaranteed to find the exact 'optimum' solution but they will typically get very close. This may seem like a significant drawback until you consider that they are used for tackling the sort of problems with which traditional and accurate optimisation algorithms cannot cope. Also for most engineering problems, getting close to the concept of the 'optimum' is sufficient as this is a rather artificial mathematical concept which is typically compromised by the needs of the real world.

There are many algorithms which come under the heading of evolutionary computing such as: genetic algorithms, genetic programming, evolutionary strategies, evolutionary programming, harmony search, ant colony algorithms, the Bee algorithm, PGSL and particle swarm analysis.

A defining feature of most of these techniques is that their workings include some processes which mimic the behaviour of some sort of natural process.

The one exception in the above list is PGSL (Raphael and Smith, 2003) which strictly speaking belongs to the wider grouping of algorithms which come under the heading of stochastic search. Also there is another closely related search technique called simulated annealing (e.g. Shea and Cagan, 1997) which has been used successfully by some people, although this only works on one solution at a time rather than a population of solutions.

For the potential user, the question arises which of the above is most suitable for my particular problem. This is almost impossible to answer. As Wolpert and MacReady (1997) show, there is no one algorithm which is better than all others for all problems. Experience has shown that generally all of these algorithms work well. The literature contains many papers which purport to show that one algorithm gives a typically small advantage in performance over one of more of the others but such comparisons are hard to sustain because they are very dependent on programming efficiency being equal for both cases and they are only valid for the chosen problems.

Therefore the approach we have chosen to use is to focus on one algorithm and explain its workings in some detail. Our choice is genetic algorithms, which are the most commonly used of all evolutionary algorithms and can be thought of as a good, robust, general purpose, problem-solving algorithm.

9.7 Genetic algorithms

In the above section, we discussed the fact that there are many types of evolutionary algorithm. There are also, likewise, many types of genetic algorithm. Space precludes coverage of all of these and so we will in this chapter concentrate on the basic techniques of the so-called canonical genetic algorithm. Having grasped the concepts of the canonical genetic algorithm, it is then fairly simple to understand the other related forms. In terms of programming, commercial software for genetic algorithms is available. For example, Matlab has an extension which allows the easy development of a genetic algorithm. However, it will be seen that genetic algorithms are relatively simple and often it is easier to write your own program because then this can be adapted to the particular problem, as required.

The basic architecture of the canonical genetic algorithm is given in Figure 9.2. As shown, the process starts with the creation of an initial population. Each

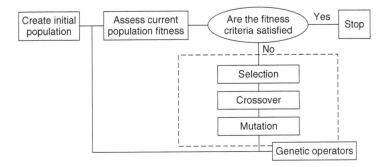

Figure 9.2 Schematic representation of genetic algorithms.

member of the population represents a potential solution to the problem being considered although, as will be shown, many features of the problem are contained within the fitness function rather than within the population.

Searching for a solution(s) using a population means that at each step, a genetic algorithm samples as many points within the search space as there are members of the population (assuming no two members are identical). This is one of the strengths of a genetic algorithm, enabling it to sample widely throughout the search space and identify areas of high performance (i.e. good solutions) on which the search can start to converge. The use of a population enables multiple high-performance areas to be identified and explored in further detail. This is part of what helps the genetic algorithm to avoid convergence on local optima.

Once the initial population is established, the genetic algorithm goes through a series of iterative processes (genetic operators in Figure 9.1) in which it adapts and modifies the members of the population based on feedback relating to how good a solution each member of the population is, until one or more good solutions are found. The judgement of how well each member of the population performs is undertaken by the fitness function. The iterative procedure continues until some appropriately chosen convergence criterion is satisfied.

The process is analogous to Darwinian evolution in that there is a population of solutions. These solutions are subject to an environment (the fitness function) which tends to favour the reproduction of the solutions which are best suited to that environment (so-called survival of the fittest). Hence solutions which suit the defined environment are evolved over a number of iterations (called generations). In the following sections, each of the features of the canonical genetic algorithm will be described in detail.

9.8 Encoding the problem

Genetic algorithms can use either a binary string or numbers to represent each member of the population (population members are described by various names including genome, chromosome, string and individual). The most common approach is to use a binary string. How this actually represents the problem to be solved is one of the challenges of implementing and genetic algorithm as there is no single best method and instead each case has to have its own bespoke representation.

Various examples will be given in the following text but the work of Hooper (1993) gives an illustration of how simple a representation can be. He devised a technique for determining the best strategy for the disposal of the sludge from a sewage treatment works using a genetic algorithm. If the sludge was to be disposed of to agriculture, then there were a given number of farms which could be used within an economical travelling distance of the treatment works. The encoding used for this problem was to represent each farm by one of the characters in a binary string. In a binary string, each character is referred to as a gene. If the gene is 1, then the policy represented by that individual is that the farm will be used to dispose of sludge, if it is zero, and then the farm will not be used.

Assuming an example where there are ten farms, a possible individual would then be as follows: [1001011101].

This represents the disposal of sludge on farms one, four, six, seven, eight and ten and no disposal on the remainder. In the way that a genetic algorithm works, it must be remembered that this would be just one possible solution within a population of solutions.

An alternative to using binary encoding is to use so called real numbers. The terminology is confusing because often the numbers are integers but as this is the standard terminology within the evolutionary computation community, it will be used here.

For example, Bradshaw (1996) was asked to derive an optimum electricity generation strategy for Great Britain. The constraints were that the strategy had to meet the demand for power at all times while allowing sufficient time for maintenance of generators etc. and taking account of the fact that it is better to generate electricity as close to the demand as possible to reduce losses in transmission.

The objective of the optimisation was to reduce the damage done by the emission and subsequent deposition of acidic gases such as SO_2. In this case the encoding used was a real number representation where each generating station was represented by a number between zero and 100. If for example the number was 20, then 20 per cent of that station's maximum generating capacity would be used within a given year.

Assuming an example where there are eight generating stations (in the real example there were over forty), then an example of an individual within a population would be:

[35,72,80,41,0,66,7,29]

With 35 per cent of Station 1's capacity being used, 72 per cent of Station 2's, etc.

9.9 The choice of encoding

There are various arguments for and against using binary and real number encoding. The argument in favour of binary encoding is based upon the schema theory (Goldberg, 1989; Holland, 1975).

The basis of the schema theory is that a schema is a similarity template that defines similarities between individuals in a population (often called chromosomes), at various points within the individuals. For example consider the following example from Parmee (2001) which uses three individuals as follows:

```
0010111001010001
1011101000110100
1011111001110101
```

The schema or similarity template for these individuals is:

#01#1#100##10#0#

Where # represents 'don't care' (in effect a mismatch).
The schema theory then looks at the following:

- The length of the binary string, L, (in this case 16);
- The defining length which is the distance between the first gene which is not represented by a # and the last gene which is not represented by a # (in this case 15–2=13) and,
- The order of the schema, which is the number of characters containing a 1 or 0 (in this case 9).

Within a given member of a population, each gene can be represented by one of three characters (#, 0 or 1) and therefore the number of schemata present within the member is 3^L. If the population contains N members, then the total number of schemata within the population is ($N3^L$).

The schema theory states that having a large number of schemata within a population increases the probability that high-quality *building blocks* (i.e. areas within a chromosome that represent good solutions to the problem being considered) will be formed.

Therefore, long, binary chromosomes should give better solutions. This is linked to Holland's (1975) thoughts on implicit parallelism within genetic algorithms.

The theory further states that binary strings have a greater information carrying capacity than a real number representation and Holland presents arguments to support this but they contain assumptions about the similarity of representations that would be used when comparing real number and binary encoding that are not necessarily valid.

There are also difficulties that occur when using binary encoding. The most obvious is that there is usually a need to translate the binary representation (the genotype) into a form where it can readily be understood by human beings (the phenotype).

Also using binary can lead to excessively long chromosomes which can be difficult to handle within a computer. Finally there is the so called *Hamming cliff* problem which occurs if a binary encoding is used for numbers.

Taking a simple example where the answer to a problem is 7:

[0111] = 7 BUT
[1000] = 8 Completely different

This means that for a genetic algorithm to converge on a solution from 7 to 8, the schema would be completely different (note the problem doesn't occur between 8 and 9). This is what is called a *Hamming cliff* – in other words all the zeros have to change to ones and vice versa.

Table 9.2 Binary and grey encoding

Decimal	Binary	Grey
1	0001	0001
2	0010	0011
3	0011	0010
4	0100	0110
5	0101	0111
6	0110	0101
7	0111	0100
8	1000	1100
9	1001	1101
10	1010	1111
11	1011	1110
12	1100	1010
13	1101	1011
14	1110	1001
15	1111	1000

The problems with this can be partially overcome by using grey scale coding. A comparison of the two approaches is given in Table 9.2.

The arguments in favour of real number representation are less complex than those given above (see Parmee, 1998). One of the main advantages is that they completely avoid *Hamming cliffs* and therefore cope much better with dealing with convergence type problems. However, the main argument in favour of real number representation is that it has been proven to work well on a wide variety of problems.

9.10 Fitness assessment: the fitness function

The processes within a GA can be thought of as mimicking those in Darwin's theory of evolution. One of the features of his theory is that individuals which are better suited to a given environment have a better chance of survival and hence of passing their genes on to the next generation. It is therefore necessary within a genetic algorithm to have some sort of function that represents *the environment*. The purpose of this function is to measure how good a chromosome is as a possible solution to the problem being considered. However, for many problems, very little is known about the form of the search space. This causes difficulties for traditional optimiszation methods but genetic algorithms are able to find good solutions without needing detailed information regarding the problem being solved. Hence the performance assessment function for a genetic algorithm is not referred to as an *objective function,* as would be the case for traditional optimiszation methods, but instead, as a fitness function.

The fitness function is always problem specific so it is impossible to give a general description and hence an example will be used. The problem for which the fitness function will be presented is that shown in Figure 9.3a (Griffiths and Miles, 2003). A vertical load is to be applied uniformly to the top of a space and

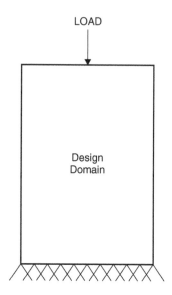

Figure 9.3a Design domain (adapted from Griffiths and Miles, 2003)

Gene Position   i = 1,2,3,4,5,...,25

Gene Allele Value   11111001 00001 00001 0011111

1	1	1	1	1
0	0	1	0	0
0	0	1	0	0
0	0	1	0	0
1	1	1	1	1

i=1	2	3	4	5
6	7	8	9	10
11	12	13	14	15
16	17	18	19	20
21	22	23	24	25

Grid with Allele Values   Grid with Gene Position

Figure 9.3b Problem encoding (adapted from Griffiths and Miles, 2003)

there is to be a support at the base of the space. In between, some form of support is needed. If the third dimension is included, this becomes a problem to find the optimum beam cross sectional area say for a simply supported beam.

The encoding of the problem is as shown in Figure 9.3b where the space is split up into squares (called voxels) and each voxel is represented as a gene in a binary string (see the upper part of Figure 9.3b). If material is present in a voxel, it is represented as a 1 and if the voxel is void it is represented as a 0. Thus in Figure 9.3b, the representation shows an I beam.

The first challenge for a genetic algorithm, given the above loading and sup-port conditions, is to generate an I beam from an initial population in which the members are created by a random number generator. This is a useful test because the correct answer is known and so it would give confidence that the algorithm could find the correct answer for loading and support conditions where the correct answer is not already known. One of the phenomena to which a beam is subjected is bending, so the fitness function should contain a function which deals with this. Therefore a cross-sectional shape is required that keeps normal stress levels, those caused by the application of a bending moment, within the allowable stress limits while utilising the minimal amount of material. Minimising the required material arrives from a desire to reduce self-weight. For this initial solution only bending moments are considered with the load being applied uniformly and symmetrically to the upper surface of the beam and the support being provided in the same manner at its lower surface. Results from a more complete approach, which also takes account of shear, are to be found in the next chapter.

The load case produces normal stresses over the entire cross-section of the beam which vary in intensity depending upon the shape of the cross section and the location of material within it. Each solution is evaluated utilising two standard stress analysis equations. The stress value σ for an individual active voxel can be calculated using the bending stress (Gere and Timoshenko, 1997):

$$\sigma = \frac{(My(i))}{I} \tag{9.2}$$

where M is the applied bending moment (Nmm), $y(i)$ is the distance of the ith active voxel from the neutral axis in millimetres, and I (mm^4) is the *second moment of area* with respect to the horizontal neutral axis.

The second moment of area for the shape (with respect to the neutral axis) is calculated as (Gere and Timoshenko, 1997) shown below:

$$I = \sum_{i=1}^{n} (y(i)^2)(A) \tag{9.3}$$

where $y(i)$ is distance of the ith voxel from the neutral axis of the shape in millimetres, A (mm^2) is the area of a voxel and n is the number of active voxels.

The neutral axis is assumed to be a horizontal line that passes through the centroid of mass of the shape (NB For more complex loading cases, this assumption is not valid). The voxel representation system applied in this research reduces the design space to a series of squares of uniform size and density. Therefore it is acceptable to calculate the neutral axis as being the average position of active voxels (average position of material).

$$Neutral\ Axis = \frac{\sum y_{base}}{Active} \tag{9.4}$$

Where *Ybase* is the distance of the material from the bottom of the design space in millimetres and *Active* is the total number of active voxels.

It is evident from a simple analysis of these equations that, for bending, the distance of an active voxel from the neutral axis is most significant in determining the optimal shape for the cross-section. Increasing the average value of y, will increase the second moment of the shape and decrease the maximum value of voxel stress.

Normal stresses require material to migrate towards the vertical extremities of the design space where the normal stress is greatest (forming two symmetrical flanges). A valid solution to this simplified mathematical model would be any shape that does not exceed the stress constraint, with the best solutions being those that satisfy this constraint with the minimum amount of material. As only normal stresses are currently being considered, there is no requirement for the genetic algorithm to develop a web (primary shear stress carrier) or any other means of connecting the two elements.

The fitness function is designed to minimise the amount of material in the beam while not exceeding the maximum allowed stress.

Table 9.3 details the constraints (must be satisfied, i.e. hard constraints) and criteria (desirable quality, i.e. soft constraints) to be satisfied by the evolved solutions. For all test cases the required criteria remain constant. Constraints are given priority over criteria through the use of penalties relative to constraint violation however, no constraint is given priority over another.

Genetic algorithms cannot deal with constraints separately as is done in many optimiszation algorithms and so the approach used is to convert them into objectives. Bearing this is mind, the forms of the fitness function is given in the following equation can be appreciated. The number 1000 is present as a multiplier to give fitness values which are in a convenient range for human comprehension and it plays no part in the actual determination of the actual fitness.

$$Fitness = 1000 \Big/ \big(Active + (1 / SVoxelMax) $$

$$+ A \times (P1) + B \times (P2) + C \times (P3) \big) \tag{9.5}$$

Table 9.3 Constraints and criteria (Griffiths and Miles, 2003)

Constraint	Variable
$SVoxelMax <= \sigma_{max}$	$P1 = (SVoxelMax - \sigma_{max})$
Criteria	Variable
Minimise material used	Active
Minimise material used	P3 = Number of exposed sides
Efficient use of material	1/SvoxelMax
Efficient use of material	P2 = (SVoxelMax - SVoxelMin)

where *Active* is the number of active voxels and *SVoxelMax* is the maximum stress value of any voxel.

$$P1 = SVoxelMax - \sigma_{max} \tag{9.6}$$
$$P2 = SVoxelMax - SVoxelMin \tag{9.7}$$

P3 = (Number Of Exposed Voxel Sides)
A,B,C = Scaling factors (A = 100, B = 30, C = 10)

σ_{max} is the maximum stress value permitted in N/mm^2 within the material and *SVoxelMin* is the minimum voxel stress.

The fitness of an individual is inversely proportional to the number of active voxels (*Active*), and hence the search favours minimum self-weight. To promote efficient use of the material, and minimise redundant stress carrying capacity, *(1/SvoxelMax)* is applied so that solutions operate at the most efficient stress values for the material utilised for the cross-section. *P1* is applied to solutions that violate the maximum stress constraint. *P1* is the only hard constraint at this point, and therefore violations are penalised heavily. *P2* again encourages the genetic algorithm to evolve solutions that operate at high stress levels to promote efficiency. *P3* is applied to minimise the development of small holes and isolated voxels by penalising solutions relative to the number of exposed voxel sides. *P3* also aids the development of solutions with minimal material as it requires voxels to form tightly packed shapes for optimality. Each isolated voxel or small hole will result in an increase in value of *P3* of four (four exposed sides), thus encouraging the GA to group together sections of material.

Scaling factors *A, B* and *C* are used to increase the weight of each variable, a standard practice when applying evolutionary searches. Scaling has been applied to prioritise the satisfaction of constraints over criteria. During initial generations the scaling factors for constraint violations are set relatively low to allow the genetic algorithm to explore the greatest range of potential solutions. During final generations the scaling factors are increased to ensure the final solution meets all constraints. Additionally, the scaling factors are applied to ensure that no single variable dominates the search. This is necessary as the range of values for each variable varies significantly. For example, during final generations, *Active* has an average value of 750 per solution, where as *P1* tends towards +/– 3. These penalties can ensure all constraints are satisfied simultaneously, without any bias. Scaling factors are also applied to the criteria elements of the fitness function to ensure no one criterion dominates another.

So what sort of results does this fitness function give? The answer depends on factors other than just the fitness function but Figure 9.4 shows typical solutions evolved by the genetic algorithm after the 2000 generations. Table 9.4 highlights the details of the best evolved solutions, at the end of a 2000 generation run (60000 evaluations, i.e. a population size of 30 chromosomes). In addition to near optimal shapes being achieved (+/– 5 voxels from known optimal), the

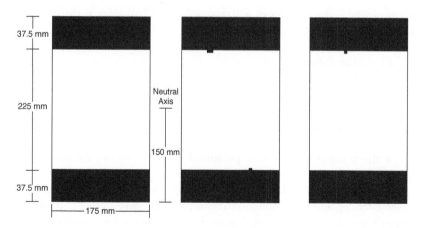

Figure 9.4 Results achieved with two-dimensional genetic operators (adapted from Griffiths and Miles, 2003).

Table 9.4 Results for the bending only problem (Griffiths and Miles,2003)

Run	Active voxels (%)	Bending stress/ N/mm²	Surface area/ exposed voxel sides	Optimal shape achieved
Initial population	50 (random generations)	Average 155.00	Average 1925	N/A
1	34.38	100.01	172	Yes
2	34.52	99.09	179	Yes
3	34.42	99.85	175	Yes

Tests conducted with a design space of 300mm by 175mm, and an applied bending moment of 200×10^6 Nm, $\sigma_{max} = 100$ MPa.

solutions are within the stress constraint limits. The genetic algorithm has a success rate of approximately 90 per cent at locating near optimal solutions.

The above example shows how, for a relatively simple problem, the fitness function can be fairly complex and also how it typically contains a mixture of objectives and constraints. However, as can be seen from Figure 9.2, there are still many other aspects of a genetic algorithm to be considered and without careful consideration of these, the above results would not have been possible.

Before we leave the subject of fitness functions, there are also problems where the fitness function is in itself a substantial computer programme. For example, if one wants to examine the thermal performance of buildings, then this is usually achieved using complex numerical analysis software (de Wilde *et al.*, 2009). In such cases the execution time of the fitness function can be prohibitive and then some form of simplification is necessary (e.g. Kean, 2004) to produce a less accurate fitness function. This can either be used for the whole search or just to approximately locate areas of high performance which are then searched in detail using the full fitness function.

9.11 Selection

Once the fitness of each chromosome (i.e. potential solution to the problem) has been calculated, the next stage is to select the chromosomes that will be placed in the mating pool and therefore be used to breed the next generation. Those individuals that are not selected will not be able to pass on their genetic information to the next generation so getting this right is an important challenge. As will be shown there is the possibility that fit individuals will not be placed in the mating pool and that unfit individuals will be chosen. This is deliberate. So long as the overall selective pressure favours the fitter individuals then from generation to generation the overall level of fitness will improve and letting some less fit individuals through is important for genetic diversity, especially during the early stages of the search.

There are various methods of selection used with genetic algorithms but, in line with the above philosophy, the basis of nearly all of them is that the chances of the fitter individuals being selected for the mating pool is stronger than that of the less fit.

The two methods that are most commonly used are roulette wheel and tournament selection, so the following looks at these.

With the roulette wheel, a conceptual wheel is constructed, with each chromosome being given a slice of the wheel which is proportionate in size to its fitness. So a fit individual is given a relatively large slice and a less fit individual a smaller slice. The wheel is then 'spun' (in reality a random number generator is used) with the result being an angle between 0 and 360 degrees. The selection operator determines which chromosome's slice corresponds to the resulting angle and that individual is then passed through to the mating pool. The wheel is spun sufficient times to provide enough individuals for mating. Note that it is entirely possible for an individual to be selected more than once.

The tournament selection method is simpler but generally more effective. In its simplest form two individuals are selected at random from the population and the fitter of the two is allowed to go forward to the mating pool. More complex forms select a greater number of individuals but the process is still the same with the fittest going forward.

With both these selection methods, it is entirely possible that the fittest individual in the population may not be selected for the mating pool. To overcome this problem, most implementations of genetic algorithms use some form of what is known as elitism where the fittest individual is automatically placed either in the mating pool or it is passed directly to the next generation. Sometimes rather than just the fittest individual, the fittest X per cent (say 10 per cent) are chosen. Some people argue that elitism is a bad thing and in such cases the usual procedure is to filter off the fittest individual from each generation and keep them on one side in a separate location so that the fittest produced by the genetic algorithm during the entire run is not lost.

9.12 Crossover

Having placed in the breeding pool the chromosomes that are to mate, the succeeding process is the creation of the next generation. There are two main operators that typically form the breeding process. The first of these is crossover (also called recombination) which is intended to mimic the processes that occur in nature when two individuals come together to create a new individual. In genetic algorithms, there are many forms of crossover, the simplest of which is single point.

The process starts with the selection of two individuals from the mating pool. For convenience, we will assume that these are both binary strings as follows:

[1000100100001]
[0111100011001]

Using a random number generator, a point is chosen some where along the length of the two chromosomes (say between the 5th and 6th bits) and the strings are then cut at this point and the new chromosomes formed by taking the left hand end of the first individual and combining it with the right hand end of the second and vice versa. The result is as follows:

[1000100011001]
[0111100100001]

Thus two new individuals are formed from the two parents. The process contains a significant degree of randomness about it. There is deliberately no attempt to determine whether or not this is a good point at which to cut each individual because although in this generation the result may not be advantageous, it may result in another generation or two in some further recombination which will give good results. The disruptiveness of crossover is both a strength and a weakness of genetic algorithms.

Crossover has the potential to take the components of two parents and combine them in such a way as to produce a much fitter individual. Conversely, it can break up a fit individual in such a way that the result is much less satisfactory. The randomness of the process is deliberate. Genetic algorithms generally deal with complex problems where it is not possible to second guess what the solution should be (if it was possible, why use a genetic algorithm) and so the ability to mix and match parts of the members of the population is part of the search process. The selective pressure, as defined by the fitness function and subsequent selection, is what ensures that the search is convergent.

However, towards the end of a search, when convergence is almost complete, crossover can be very disruptive and at this stage it can in some circumstances be advantageous to stop the genetic algorithm and instead apply a more traditional, deterministic algorithm such as hill-climbing.

There are many other forms of crossover. The basis of most of these is the use of a mask as shown below in Figure 9.5. This shows how a mask can be used for one

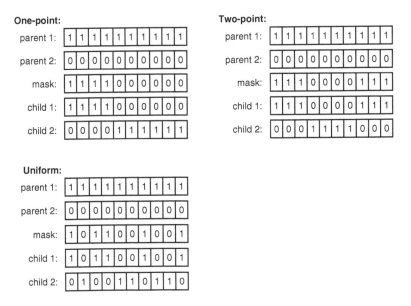

Figure 9.5 Mask-based crossover.

point, two point and multi-point crossover. In effect the mask is another binary string of the same length as the chromosome. In each position on the mask is either a 1 or a 0. If there is a 1, then that part of the chromosome remains unchanged. If there is a 0, then the corresponding bit from the other parent is imported into the chromosome. This gives the basis of how crossover works but as will be show later, in the section on topological reasoning, the choice of what mechanism to use can have an impact on the ability of the algorithm to search effectively.

9.13 Mutation

The other main breeding operator used with genetic algorithms is mutation. In nature, mutation allows species to change their characteristics over time and hence introduce features that were not present in the original population. In genetic algorithms, the working of mutation is very similar. Within the original population, there is only a limited amount of genetic information and while crossover can combine and mix this information in new ways to try and find desirable results, it is possible that from time to time, the introduction of new material will be useful. This is the basic thinking behind the mutation operator.

Mutation has to be used carefully but its impact can be highly beneficial. If there is no mutation, then often the result is that the genetic algorithm fails to find the best areas within the search space. However, if too much mutation is allowed, then its impact can become disruptive and the process can degenerate into a random search.

Table 9.5 An example of mutation

Old chromosome	Random numbers				New bit	New chromosome			
1 0 1 0	0.801	0.102	0.266	0.373	–	1 0 1 0
1 1 1 0	0.120	0.096	0.005	0.840	0	1 1 0 0
0 0 1 0	0.760	0.473	0.894	0.001	1	0 0 1 1

The typical method of applying mutation for a binary chromosome is to define a so-called mutation rate and then generate a random number between zero and one for each bit. If, for a given bit, the number is less than the mutation rate, which is typically about 0.01, then the bit is 'flipped'; that is a 1 is changed to a 0 or vice versa. To show how this works an example is given below in table 9.5.

The population in the table contains 3 chromosomes with a string length of 4. The original chromosomes are given on the left hand side. The mutation rate in this case is set at 0.008. The random numbers for each position on the strings are given in the centre of the table. As can be seen only two are less than 0.008 so only two bit flips occur. The amended population is given on the right.

With real number chromosomes, mutation is slightly more complex but works on similar principles. Typically two types of mutation operator are used with real numbers, jump and creep. The former allows a given number in a chromosome to change to any value within the allowed range and the latter just makes a small change. The way that this is then applied is to initially generate a random number between one and zero for each member of the chromosome and if the number is below the mutation rate, the member is selected for mutation. A further random number between one and zero is then generated for each member. If the number if below say 0.5, then the jump operator is used and if it is equal to or above 0.5, then the creep operator is used.

Referring to the work of Bradshaw (1996) on power stations, for each member of a chromosome, the possible range was between zero and 100 per cent. In practice the upper limit was set at 80 per cent to allow for maintenance time. So the jump operator allowed the number to be mutated to any value between 0 and 80. The value to be applied was again determined by using a random number generator. The creep operator allowed the number to be changed by ±5 per cent, again with the value being determined randomly. For this particular problem, it was found to be advantageous to apply the mutation operators equally at the start of the search process (i.e. below 0.5 use the jump operator and equal to or above 0.5 use the creep). However as the search progressed, the jump operator was found to be too disruptive when applied at a high rate and so the rates were progressively changed as the search progressed, so that in the later stages, the jump operator was rarely applied.

Table 9.6 Example of Inversion

Prior to Inversion	After Inversion
[101010101010]	[010101010101]

9.14 Inversion

There is another operator that is sometimes used. In this the entire order of the chromosome is swapped so that the right hand end becomes the left hand end etc. See Figure 9.6 below for an example. In practice, inversion is rarely used because it is generally found to be too disruptive.

9.15 Convergence and results

For most applications of genetic algorithms, information about the likely value of the final answer to the problem is completely lacking (if it is available, why use a genetic algorithm?) and therefore it is not possible to say, with absolute certainty, when the search process is complete. In such circumstances, it is usual to let the search progress for a pre-determined number of generations. However, some means of checking that the answer obtained is probably somewhere near the best possible is necessary. One way of trying to assess whether or not this has occurred is to plot a graph of the change in the fitness that occurs as the search progresses. Typically the values plotted are for the fittest individual within a population, although it is also usual to plot the average fitness of the population as well. An example plot of the fittest individual is given below in Figure 9.6. In this particular case, the correct answer to the problem is known and it is given a fitness of 100. However, this approach can give false confidence if the algorithm has converged on a local optimum. For this reason, it is good practice to run a genetic algorithm several times with the same input data. The inherent randomness will ensure that the process used to get to convergence is different each time and so, hopefully, for at least one of these runs, the algorithm will avoid convergence on sub-optimal solutions. Generally genetic algorithms can be relied upon to find a solution which is very close to the optimum if not the exact optimum.

As can be seen from the figure, the process gets close to the best answer sometime around the 300th generation and then slowly converges so that by the

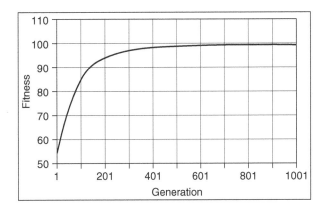

Figure 9.6 Example of change in fitness with generation.

600th generation it finds the best answer. This is for a case where the answer is known, but for examples where the answer is not known, one would hope that the change in fitness will follow a similar path although to some extent this is a function of the problem space.

In cases where the problem space is reasonably continuous, then one would expect the behaviour to be similar to that shown in Figure 9.6. However, even when the process follows a reasonably smooth curve, it is not possible to guarantee that the answer obtained is a good one and sometimes a sudden jump occurs in the fitness curve. This for example could be when mutation takes the search to a different area of higher performance.

Where the problem space contains significant discontinuities, then one would expect a plot of fitness against generation to be much less smooth than that shown in Figure 9.7. In such cases, it is much more difficult to judge with confidence when the process seems to have converged.

In many cases with genetic algorithms, the code is arranged in such a way that just 'the best' result is presented to the user. This really does not make full use of the powers of genetic algorithms. Probably the most comprehensive method for location and presenting areas of high performance has been developed by Parmee (1998). His approach has been specially developed for design problems where rather than finding 'the best' answer, one is more concerned with learning about the problem space and finding areas of high performance. The details of Parmee's method are too complex to be explained here but the overall concept is that at each generation, chromosomes that pass a pre-determined level of fitness are copied to a separate area where they are stored and at the end of the search process, the contents of the area are presented to the user using a variety of techniques including box plots and hyper planes. The plots can be presented in a variety of formats including variable space and objective space plus the fitness landscape. The approach is sophisticated and powerful and is recommended to anyone who wishes to progress beyond the basic techniques that are being discussed here. A modification of this approach is contained in de Wilde *et al.* (2009).

There is also a simpler and less powerful technique that is sometime employed using Pareto fronts. A good example of the application of this to structural engineering is to be found in the work of Khajehpour and Grierson (1999) and Grierson and Khajehpour (2002).

9.16 Conclusion

The above covers the basic techniques for developing a so-called canonical genetic algorithm. There are many variations on the techniques described but for getting started and learning about genetic algorithms, it is recommended that initial work uses the above approaches and then develops further complexity as necessary. The additional techniques are to be found in the literature. Also, there are other very useful algorithms in addition to genetic algorithms and again once the basics of evolutionary computation have been mastered, it can often be beneficial to try other procedures.

10 Application of genetic algorithms for design

Building design; BGRID representation; BGRID: reproduction, crossover, mutation; The BGRID fitness function; BGRID: a typical user session; Creating the initial population; Component sizing; Controlling the search; Search results; BGRID evaluation; Further developments; Evolutionary operators; OBGRID and orthogonal buildings; Evolutionary operators; Illustrative example: an orthogonal building; Results; Conclusion.

10.1 Building design

Design is a complex, multi-faceted, multi-disciplinary activity and so it is not possible to present a one solution fits all problems form of genetic algorithm. Instead, this chapter will give an introduction to how one can deal with a specific design challenge and look at how this can be tackled using genetic algorithms. The focus of the chapter is the conceptual design of typical, framed commercial buildings such as multi-storey office blocks. These are an interesting challenge because they have a high degree of input from the major design disciplines of Architecture, structural Engineering and building services. Fully incorporating all aspects of the design decision-making process for each discipline is beyond the abilities of the current technology due to the limitations of the known forms of representation. Nevertheless, as is shown, even in its current form, the technology offers a useful and powerful search tool.

A typical, framed commercial building such as an office, a hospital or a hotel consists of columns and beams with the floors being either concrete or composite slabs. For the external skin of the building some sort of infill is used, for example bricks or block work. The columns and beams of the frame are usually either steel or concrete, although increasingly timber is being used because of its lower environmental impact.

The design of such a building is undertaken by a multi-disciplinary team of designers. The disciplines involved typically include architects, structural engineering and building services engineers. The team is assembled at the request of the client who is the person or organisation which is commissioning the building. The client produces a brief which is interpreted and expanded by the designers. The brief is in effect a list of the client's requirements in terms of the style, performance and functionality of the building. The designer's task is to understand the client's aims and objectives and to convert these into a description of a building which meets these. The design is then taken on by a contractor and sub-contractors and it is their task to convert the description into an actual building. Sometimes contractors are also involved in the design stages because their expertise can be usefully exploited so that the result is a design which can be easily and economically constructed.

Design can be thought of as consisting of a number of stages, although in reality it is more of a continuous process. As mentioned above, the very earliest stages go by various names but typically are called client briefing or brief development. Briefing starts with the client expressing what sort of building they want, how they want the building to perform in terms of things such as utility and energy usage, and what their overall parameters are in terms of cost and the areas required. There then follows a series of conversations between the client and one or more of the designer disciplines to develop these basic ideas into a more comprehensive statement of what is required. At this stage, typically there are no drawings other than a plan of the area in which the building is to be located (i.e. a site plan).

The next stage is typically referred to as conceptual design, although other names such as embodiment are also used (Pahl and Beitz, 1988). It is at this stage that the fundamental decisions regarding the form of the building are taken and typically by the end of this process, 80 per cent of the costs of the building are fixed. Therefore it is vital that the right decisions are made otherwise the result can either be that the client incurs unnecessary costs or obtains a building which does not satisfy their requirements. Generally during conceptual design, the details of the design have yet to be fixed and so often decision making is based on rough estimations and heuristics rather than absolute values.

Following conceptual design, there then follow a series of stages in which detail is gradually built up. This involves both graphical descriptions (i.e. CAD drawings or more recently building information models referred to as BIMs) and for the more technical disciplines such as structural and building services engineering, calculations are undertaken to model and assess the performance of various components. The calculations typically involve complex numerical analysis techniques such as finite elements. For the final stages the detailed drawings or BIM and other contract documents are produced. Also there is a growing trend to produce a process model as well as a static graphical description, the latter helping to control the construction phases of the building (Richter *et al.*, 2009).

In this chapter, the focus is on the conceptual design stage. It will be shown that conceptual design is a complex and demanding task which needs to be supported by appropriate computational techniques. However before this discussion starts, one very important ground rule has to be established. In all complex decision-making processes, it is vital that the decisions are made by human beings, not computers. Computers are excellent at processing and handling large amounts of information but they do not possess real world knowledge (other than that which has been specifically encoded within software) and they therefore lack common sense and judgement. So, in any decision making process where a human uses some form of computational support, the latter has to be used appropriately. If this is done the result will be far superior to that which can be produced by other techniques. Therefore it is vital that the process makes the best use of the strengths of both the human being and the computer.

Typically during the conceptual design of a building, each design discipline evaluates several options. For example, the structural designer will probably consider steel and concrete options and probably one or two variations of each of these. So typically each design discipline may look at as many as six possible options. Khajehpour and Grierson (2003) estimate that for the design of a typical building, at the conceptual design stage, there are at least 400,000 feasible options from which the design team can choose (typically the number of options is much greater (Miles *et al.*, 2007)), so if each designer looks at six options, there is a huge number of possibilities that have been ignored. Of these 400,000, it is possible that some will be poor choices that any experienced designer would automatically reject. Also some may be very similar but nevertheless, the evidence is overwhelming that unless they are incredibly lucky, the design team will manage to do no more than produce a design solution that satisfies the constraints, a so-called satisficing solution.

Therefore the design team need assistance to help them search through the multiplicity of feasible solutions to find the best from the range of possible options and it is here that genetic algorithms and other search algorithms can be of use. To show how a relatively simple search algorithm can be used with the design process, an example follows. This is based on the work of Sisk (1999) and it involves a system she developed called BGRID. The domain of BGRID is limited to buildings with steel frames. It was always intended to extend the work to include concrete framed buildings but lack of time prevented this. Nevertheless, the system is a good example of a simple design search tool. The discussion will first look at a technical description of BGRID before then taking the reader through a typical user session.

10.2 BGRID representation

In a genetic algorithm the representation is the choice of features of the problem that are to be included in the chromosome and the encoding method (real number or binary). It is usually advantageous to keep the representation as simple as possible. In BGRID, the chromosome contains four types of information:

- *The (x,y) coordinates of each column centre.* As the number of columns varies between solutions this means that BGRID does not have a fixed length of chromosome.
- *The structure-services integration strategy.* This describes how the services (i.e. ventilation, plumbing, electricity, etc.) are incorporated with the structural elements with it being assumed there will be a suspended ceiling. There

are three structure-services integration options within BGRID, these being, separate (i.e. all services below the structural frame supporting the floor), partially integrated (i.e. the services being partially intertwined with the structure) and fully integrated. The option chosen affects the grid because for each option there is a limited number of structural solutions, which in turn dictates the economical column spacing. The choice of structural system influences the floor to floor height and hence has a major impact on cost because the taller the building, the greater the cost.

- *The environmental strategy.* There are three options within the system, natural ventilation, mechanical ventilation and air conditioning. The choice of these is influenced by factors such as the building location (e.g. in a city centre where air quality will be poor, natural ventilation would not be a good option), the floor to ceiling height and building depth (e.g. the distance between windows on opposite sides of the building) and they have a major impact on the height of a building, its floor plan and hence its cost and utility.

- *The final component of the genotype is the clear floor to ceiling height.* As described above, the overall height of a building has significant cost implications and also dictates whether or not it can be illuminated using natural daylight and naturally ventilated.

Real number coding is used in BGRID. It would have been perfectly feasible to use binary but the resulting genome would have been relatively long and there is always the inconvenience of having to translate from binary to real number every time one wants to look at the numbers within the genome. The spacing of the column grid in plan is represented by integers, one for the X and one for the Y coordinate of each column. The structural services and environmental strategies are likewise represented by integers and the floor to ceiling height is a real number. A rather short example of a genome is given below in Figure 10.1. The left hand part (shaded in dark grey), contains the X column coordinates, the middle part (shaded light grey), contains the Y column coordinates and the right had end contains, the other parts of the genotype.

10.3 BGRID: reproduction, crossover, mutation

The elitism policy used in BGRID is that only the fittest member from a given generation is automatically passed through to the next generation. This ensures that the fittest solution obtained is not lost. The choice of how much genetic material should be passed directly through to the next generation is problem-dependent and should be made based on tests to find what gives the best over-

| 0 | 25 | 50 | 75 | 90 | 100 | 0 | 20 | 40 | 65 | 80 | 90 | 0 | 1 | 2.9 |

Figure 10.1 An example genotype.

all performance. Within BGRID roulette wheel selection is used to determine which chromosomes are to be passed through to the mating pool (if the work was to be repeated, it is probable that another selection method would be used).

The crossover mechanism has to take account of the structure of the genome which, as can be seen in Figure 10.1, is split into three distinct segments. To avoid creating a lot of illegal solutions during crossover, the operator is applied separately to each of the three parts of the string. Single point crossover is used for each part, with the actual location within the string being chosen randomly. The choice of single point was made for BGRID because more complex schemes, such as multi-point tend to produce a lot of illegal solutions and also because the chromosome length is not fixed.

If crossover produces column spacings which do not fit the imposed constraints (e.g. there are planning grid constraints based on required work spaces within the building), the nearest suitable value is determined and the illegal value altered to conform. Where crossover produces a chromosome with illegal column spacings (i.e. not all the values increase from left to right or top to bottom), the ordering is adjusted to suit. Also a further check is run to check the column spacings. Spacings which exceed 18m (the maximum spacing allowed by the structural systems used in BGRID) are reduced by inserting another column. Where columns are too close, one column is removed or the spacings are adjusted.

BGRID's mutation operator uses a mutation rate of 0.01. For the genes which contain the column coordinates the range of mutation is restricted. How this works is best explained by using an example as follows:

> For the gene selected for mutation, the first step is to look at the values of the two genes on either side of it (obviously for genes at the end of a segment one can only look in one direction). So taking for example the chromosome in Figure 10.1, if the third gene with a value of 50 is the one that has been selected for mutation, it is necessary to ensure that the resultant value is consistent with the various grids that have been defined for the building. So the mutation process starts by setting up a range of possible values which are consistent with the grid-spacing constraints and fit in between the two adjacent column spacings (in this case 25 and 75m). The new value is then chosen randomly from within the range of values.

For the third part of the genome, the mutation mechanisms are somewhat different. For the structural services, the integer values are:

0 Separate,
1 Partially integrated,
2 Fully integrated.

So if the existing value was zero, the mutation process chooses randomly between the other two strategies. Likewise the environmental strategy consists of three integers so the mutation mechanism is identical. The floor to ceiling height is chosen from within a range between 2.6 and 3.3m.

10.4 The BGRID fitness function

The development process used for BGRID involved close collaboration with practising designers. At an early stage of this process, they specified that they required a fitness function which would allow them to vary their search of the design space. They wanted this so that they could assess how their various choices impacted on the solution which was obtained. Such features within a design decision support system allow the user to operate the system in a flexible manner, facilitating a full exploration of the design space while leaving the final decision to the designer. As discussed above, this is a vital feature of any design search software.

Within the BGRID fitness function, this flexibility is achieved by allowing the user to alter the weights of the individual components of the function. This feature also allows the different types of users (e.g. architects, structural engineers, etc.) to alter the bias of the search to suit their own discipline-specific viewpoint.

The fitness function initially checks whether or not the hard constraints have been violated. Any member of the population in which the hard constraints are not satisfied will incur a solution in the form of a reduced fitness. This will reduce its chances of being selected for breeding. The hard constraint checks include:

- Checking if the overall height of the building is less than the specified maximum value (if present),
- Checking if the design option is compatible with the chosen structural system, particularly with respect to maximum and minimum span lengths between adjacent columns,
- Check the uniformity of grid, this being defined as the difference between the maximum and minimum span length (this has a significant influence on buildability). There are user-defined limits for this.

If the above are not satisfied a penalty function is applied to the individual's fitness. The value of penalty varies for each constraint. The fitness of the individual design solution is multiplied by the inverse of the penalty function with the penalties being as follows.

Height:	$PF1 = 1$	*if height > height restriction*
	$PF1 = 0$	*if height <= height restriction*
Design Option Suitability:	$PF2 = PF2 + 0.25$	(10.1)
Uniformity:	$PF3 = PF3 + 0.25$	(10.2)

As can be seen:

1 For the first penalty function, $PF1$, its value is 1 if the constraint is violated (i.e. the overall height is greater than the height restriction) and otherwise it has a value of zero.

2   For the second penalty function, *PF2*, the value is initially set to 0 for each individual and then increased by increments of 0.25 for each bay (i.e. spacing between adjacent columns) that doesn't fall within the economic span range of the given structural design option.

3   The third penalty function, *PF3*, also increases by increments of 0.25 for each different bay dimension, (e.g. if there are three bay sizes of say 9, 12 and 15m within the grid, a penalty function of 0.5 will be applied. If all the bay sizes are the same, then the penalty function is zero. This means the system favours solutions with equal column spacings).

The three penalty functions are then added thus:

$$PF = \sum (PF1 + PF2 + PF3) + 1 \qquad (10.3)$$

The second part of the fitness function deals with the soft constraints. Soft constraints are more difficult to quantify and the way this is dealt with within BGRID is that each individual is assessed relative to all the other individuals within the population. That is, the worst and the best examples of each criterion are used to determine how good the design solution is for this particular aspect of the design. This is achieved as follows:

$$fiobj = \frac{(fi - fibad)}{(figood - fibad)} \qquad (10.4)$$

Where *fibad* = fitness value of worst individual; *figood* = fitness value of best individual; *fi* = fitness value of evaluated parameter

The objective is to maximise the value of *fiobj*. The process is iterative with usually two passes through the population being sufficient to give a reasonably accurate answer.

There are three soft constraints, each of which is weighted by the user to reflect their importance for a particular search strategy. This is the main mechanism by which the user influences the search. The weight factors range from 0 to 4, with 0 being *unimportant* and 4 being *highly significant*.

The three components are:

• Large clear span;
• Minimising cost;
• Minimising environmental damage.

The *large clear span* component enables the user to search for solutions with as large a span as is feasible, within the constraints of the design. Often clients prefer buildings with as much column-free space as possible because of the greater flexibility of such configurations but larger spans create deeper beams, therefore the building is taller and hence the cost is greater so this option can be relatively expensive.

The *minimum cost* component within the system doesn't actually optimise on real costs but instead uses features which are cost-significant, these being :

- Total weight per floor area (kg/m^2) (*Including steel, slab, deck, reinforcement, services*);
- Overall building height (m);
- Net/gross floor area ratio.

The aim is to minimise the first two factors and maximise the third. It was felt by the designers who collaborated in the development of BGRID that such an approach, rather than using material quantities and costs, was easier in terms of the amount of calculation required and sufficiently accurate, for the conceptual design stage, to guide the search towards low-cost solutions.

A similar approach is used for the *minimising environmental damage* component with the following aspects being considered:

- Depth of space;
- Clear floor-to-ceiling height;
- Location.

The *depth of space* is defined as the distance between opposite windows (making allowances for atria etc.) A higher than normal floor to ceiling height (usually > 2.9m) is required if natural daylight and natural ventilation are to be viable options but if a building is situated in a city centre the only ventilation strategy that is viable is air-conditioning because of the poor air quality.

The above formulae are summed to give an overall fitness as follows:

$$F_i = \frac{\sum(fiobj * wf)}{\sum(PF) + \sum(wf)} \tag{10.5}$$

where *fiobj* = sub-fitness of individual for each evaluating parameter (from eqn 10.4); *wf* = weight factor determined by user for that evaluating parameter; *PF* = penalty function for the hard constraints

This determines the raw fitness of each individual in the population and is used to determine the rank order. The Standard Fitness Method (Bradshaw and Miles, 1997) is then used to assign pre-determined fitness values which effectively determines the degree of selective pressure assigned to each individual. The pre-determined fitnesses are calculated using an algorithm based on the normal distribution and are then used to determine fixed slot sizes on a roulette wheel. The individual with the largest rank order (i.e. raw fitness) is then given the largest slot, the next in rank order the next largest etc. As mentioned above, if this work was to be repeated, it is probably that a more efficient selection method such as tournament selection would be used.

10.5 BGRID: a typical user session

BGRID starts with a series of screens which require the input, by the user, of information about the building which is to be designed. The data input is divided into four sections:

1   Geometrical information (plan dimensions and number of floors). The design is restricted to rectangular floor plans (more complex shapes are dealt with later). The user is asked for the overall X and Y dimensions of the building and the number of storeys. During the search process, BGRID will almost certainly produce some solutions which do not exactly conform to the required plan dimensions but which are high fitness solutions. These are not penalised by the fitness function as it is assumed the designer would wish to be aware of them.

2   Site location. The system offers three options, *urban, suburban and rural*. Also the user is asked to specify any overall height restrictions on the building (usually imposed as a planning requirement) and the maximum desirable floor to ceiling height. The above factors have a significant bearing on the available options for lighting and heating and ventilation.

3   Location of architectural spaces, e.g. cores and atria (Figure 10.2). The user is first presented with a screen which asks for the maximum occupancy (in terms of number of people) of each storey and from this a minimum width of fire escape is estimated using heuristics. The process then moves on to fixing the number and location of cores and atria (if any). In the search process, BGRID may, in a given individual, make slight adjustments to the sizes of cores and more often to atria, to better fit them to the structural grid.

4   Dimensional constraints (see below)

The dimensional constraints control the grids that can be generated in the initial population. The user is asked to input the various grid dimensions, which are to be used by the system. The user chooses from a menu which contains all the available grid sizes and these are then used within BGRID as follows:

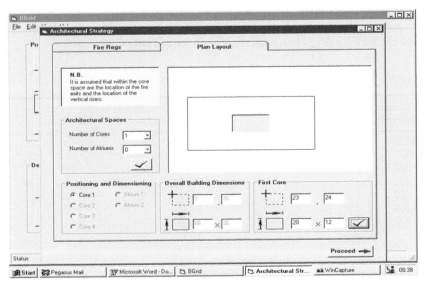

Figure 10.2 Specification of core and atria spaces.

The building dimensions are built up in multiples of the initial modular (i.e. constructional) grid. The design of a building should consider apparently insignificant elements, such as the length of fluorescent tubes and tile dimensions, as their cumulative impact can have a significant effect on the layout of the building. Based on advice from the designers who collaborated in the BGRID development, the user is offered a choice of initial modular grid dimensions of 1200, 1500 and 1800 mm.

The next dimension is the critical grid, this being the absolute minimum distance between two adjacent columns. This is determined by the intended use of the building. Again, based on advice from the industrial collaborators the available critical grid dimensions within the system are 2400, 3000 and 3600 mm.

Following this, the user inputs the preferred minimum bay dimension (i.e. the preferred distance between two adjacent columns). Users are also given the opportunity to specify the maximum preferred distance between two adjacent columns. These constraints are then used to limit the form of the grids generated in the initial population. This results in a massive reduction in the search space, which is both an advantage and a disadvantage. The benefit is that the problem is much less complicated to solve but occasionally the search space is so constrained that the search is unable to find solutions which are significantly better than the best of the initial (i.e. random) population. The available minimum and maximum bay dimensions within BGRID are given in Table 10.1. Again these have been provided by the industrial collaborators.

The next part of the system allows access to all the background information, such as the various structural configurations and section sizes used by the system to generate the design solutions. Various default sizes are used within the system but the user is free to change all these, as described below.

Table 10.1 Bay dimensions

Available minimum bay dimensions (mm)	Available maximum bay dimensions (mm)
4500	6000
4800	7200
5000	8000
5400	8400
600	9000
7200	10000
8400	10800
–	12000
–	13200
–	14400
–	15000
–	18000

10.6 Creating the initial population

In the initial population for each individual the bay dimensions are chosen randomly from the ranges of available values. The other parameters are likewise generated randomly using values from the available pre-determined ranges.

10.7 Component sizing

BGRID sizes structural components using span/depth ratios. This level of detail was felt by the industrial collaborators to be adequate for conceptual design, although subsequent work in this area (see below) has used a far more accurate approach. The user is free to override any of the system's decisions regarding section sizes and span depth ratios and they can, if wished, insert their own values. The provision of facilities that enable the user to access and adjust the assumptions and defaults within BGRID is a deliberate strategy. The collaborators asked for a system which was both transparent and flexible. A typical example of the information provided for a given structural flooring system is shown below in Figure 10.3.

Currently BGRID contains information regarding three types of flooring system. For the short spans there is the Slimflor system (Slimflor is a trademark of CORUS plc.). In this system, the structural and services zones are fully separated. The Slimflor beam is integrated within the deck, thus minimising the depth of the structural zone. This is advantageous for a highly serviced building, allowing complete freedom for the horizontal distribution of services. Each of

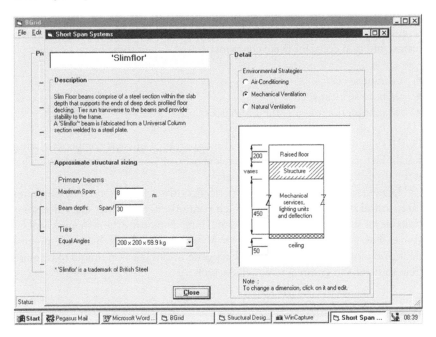

Figure 10.3 Short span structural system background information.

the environmental strategies available within BGRID requires different depths of zones for items such as air conditioning and ventilation and BGRID contains suitable defaults for each strategy. The user is free to change any of these default depths if this is necessary. For the Slimflor system, the ties, which provide frame stability, can be selected from a list of angles provided within BGRID. These dimensions are then used to calculate the overall height of the building. With all the flooring systems, BGRID contains values of minimum and maximum allowable spans.

For the medium span system, a composite steel beam and concrete slab system is used with the structural and environmental zones being partially integrated. The default vertical dimensions for each ventilation strategy are again contained within the system and can be changed if required.

For the long span system, a steel stub girder and composite slab system is used. The maximum span for this system is 18–20 metres. Three typical grids exist and these are fully documented within BGRID. The grid descriptions include details of the bottom chord, secondary beam and overall depth of the structure. Again, the user is free to change all of these values. For the long span system, the services are fully integrated within the structural zone because of the use of castellated beams. Further information on the flooring systems is given in Sisk (1999).

10.8 Controlling the search

The next part of the user interaction concerns the search process and how to guide this by altering the fitness function weights. As described above, the fitness function contains three components, the aim of these being to:

- Minimise cost;
- Maximise clear span;
- Maximise the use of natural resources.

BGRID requires the user to choose the importance of each component using weight factors which range from 0 (irrelevant) to 4 (extremely important). This is the final stage of BGRID's user input.

The next step is to activate the genetic algorithm, which randomly generates three initial populations, each of which contains 50 solutions. The three solutions are for short spans, medium spans and long spans using the systems described above. The genetic algorithm runs for 50 generations with the whole process taking no more than a few seconds. Fifty generations has been found to be more than sufficient to ensure a reasonable search.

10.9 Search results

Once the search is completed, the next stage is for the user to evaluate the results. This is done for a number of reasons:

1   The choice of search criteria may have produced results which are not what is expected or desired. In this case, the user can easily run the system again using different search criteria.

2   The user may wish to explore a range of options to learn about how changing the objectives influences the results which are obtained. As above, the system can be run again.

3   In all cases, the results need to be checked to see that they are appropriate and sensible. As discussed above, computers lack real world knowledge and common sense and although steps can be taken to incorporate aspects of this, the results can never be truly comprehensive.

The user is able to access a range of information including the maximum, average and minimum fitnesses for each generation. Details of the 'best' solution for each generation are provided in both textual (Figure 10.4) and graphical form, the latter showing a floor plan with column spacings and positions of cores and atria (if present). Also, the system provides the best design solution in both graphical and numerical form for each of the short, medium and long span structural systems. The planning grid, the structural grid, the vertical dimensions, overall height of building, weight of steel and total weight of floors are provided, as is the net/gross floor area, the wall/floor area ratio, the environmental strategy and the fitness.

The user is able to edit any of the system's solutions, for example moving the position of the cores or adding an atrium. BGRID automatically checks any amendments to ensure that none of them violates the constraints and re-evaluates the design working out a new fitness value. This allows the user to see how beneficial their changes are and also to check that the search has been effective.

10.10 BGRID evaluation

All computer software needs to be comprehensively checked. With software where the range of answers is relatively restricted this is easy but for search engines which are dealing with huge search spaces it is not possible to do an exhaustive search of a complete domain to ensure that the system is working correctly. Genetic algorithm software can be tested on relatively simple problems where the answer is known but there also needs to be further evaluation on domain relevant problems. There is also the further challenge of ensuring that the software is useful and acceptable to potential users.

BGRID has been developed in collaboration with practising designers and has been evaluated throughout its development, by architects, structural engineers and services engineers. Although in the earlier stages some teething problems were identified, in general, and especially towards the end of the development process, the response to BGRID has been very positive. Evaluators appreciate the flexible way in which BGRID can be used to investigate design problems. Also the level of detail, both in terms of the design and the underlying

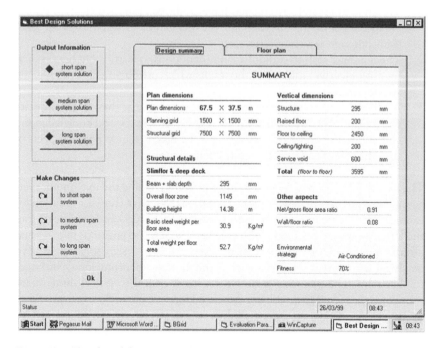

Figure 10.4 Text-based design summary.

calculations, was deemed to be correct for conceptual design, although as stated above, further work (see below) has used more accurate approaches. One of the advantages of using computationally assisted conceptual design is that the level of accuracy can be increased, thus reducing risk.

10.11 Further developments

One feature of BGRID that is very limiting is that it can only cope with buildings with rectangular floor plans. This deficiency has been looked at by Shaw *et al.* (2005a), who developed a genetic algorithm-based method for designing buildings with orthogonal boundaries. This further development of BGRID is called OBGRID.

In terms of data requirements, the main difference between BGRID and OBGRID is that for the latter, there is more effort in defining the building floor plan. As the shape is more complex than just a single rectangle, this is a somewhat more involved procedure for OBGRID. However as OBGRID breaks all areas down into rectangles, the discussion will start by looking at the differences between BGRID and OBGRID in dealing with simple rectangles before moving on to look at a more complex shape.

The basic chromosome for a rectangle is identical to that used for BGRID (see Figures 10.5 and 10.1) with the first part containing the X column spacings, the second the Y column spacings and the third part the structural system, services integration/environmental strategies and the floor to ceiling height.

Unlike BGRID, no effort is made to constrain column positions to realistic spacings. This is to make the GA search for solutions in both the feasible and infeasible regions and hence improve the search.

10.12 Evolutionary operators

There are some differences between BGRID and OBGRID in terms of the evolutionary operators. The differences in terms of a single rectangle are explained here:

- *Mutation*: If the mutation operator selects a gene from sections 1 or 2 of the genome (Figure 10.1), the gene is replaced with a randomly generated value between 0 and the limits of the floor plan. Unlike BGRID, which restricts the new spacing to a value between the two adjacent genes, OBGRID simply generates a random spacing and, if required, sorts the genome so that the column spacings increase from left to right. If a gene from section 3 (Figure 10.1) is chosen, it is mutated as normal.
- *Crossover*: This is the same as for BGRID.
- *Selection*: Tournament selection is used rather than roulette wheel (Goldberg, 1989).

The basic form of the fitness function is the same as that used in BGRID although OBGRID uses a quadratic penalty function (Richardson *et al.*, 1989). This relates the size of the penalty to the degree of transgression, with the greater errors being more heavily penalised (Shaw *et al.*, 2005b).

10.13 OBGRID and orthogonal buildings

In this section, the additional techniques needed for dealing with orthogonal building floor plans are descbribed. The basic procedure is that the orthogonal floor plan is partitioned into rectangles which can then be processed using a modified version of the rectangular methodology described above. The polygon partitioning is achieved using a 'sweep line' algorithm (Shamos, 1978). This moves an imaginary line, the 'sweep line', over a polygon from top to bottom or left to right. During a sweep, the line is stopped at 'event points' when the polygon is partitioned (O'Rourke, 1998). For OBGRID, event points are any reflex vertex on the boundary (Figure 10.5).

Partitioning is completed in two stages:

- *First stage*: a line is swept from top to bottom. When the line encounters an event point, it extends the relevant boundary edge horizontally across the floor plan until it encounters another edge and splits it, at the point

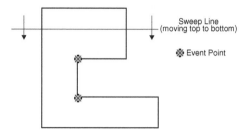

Figure 10.5 An example sweep line.

of intersection. This partitions the building into several, 'thin' rectangles (Figure 10.6a).

- *Second stage*: a line is swept from left to right across the boundary, further partitioning the rectangles created by the first stage: creating a grid pattern (Figure 10.6b). For each floor plan, there is a unique partitioning.

An adjacency graph is used next to tell the genetic algorithm which sections of the building are connected to one another. This is used later to ensure column spacing continuity between adjacent rectangles (Figure 10.7).

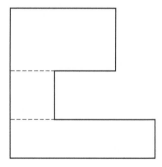

Figure 10.6a Stage 1 of partitioning.

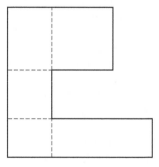

Figure 10.6b Stage 2 of partitioning.

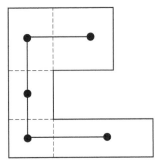

Figure 10.7 Adjacency graph.

The graph associates a node with each partitioned section and links it to an adjacent section create the adjacency graph. With the floor plan decomposed into a grid of rectangles, each partition must now share at least one edge with another partition (with an upper limit of four).

With the building being partitioned, a genome can be generated for each rectangle. The initialisation process starts by selecting the left most upper partition (this is an arbitrary selection as the initialisation process could theoretically start at any partition).

Having initialised the first partition, the algorithm selects an adjacent partition and generates a new genome for it. However as the next partition must share a common edge, the algorithm firstly copies the column spacings for this edge. For example, in Figure 10.8a the x and y spacings from section 1 are copied into the adjacent partitions. New spacings are then generated for the remaining directions (Figure 10.8b). In complicated buildings it is possible that both directions have been initialised, in this instance the partition simply copies spacings from adjacent sections.

By constantly maintaining and updating the status of neighbouring sections, using the adjacency graph, the algorithm ensures continuity of column spacings throughout the building. This continuity is vital to prevent the building from becoming a series of blocks that when placed together do not form a coherent

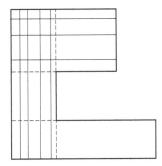

Figure 10.8a First partitioning.

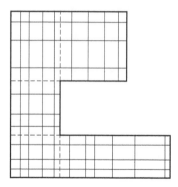

Figure 10.8b Final partitioning.

solution. For example, in Figure 10.9 when considered in isolation each section is valid however, when considered as a whole, the building's layout is flawed because the columns do not align.

The third section of the genome is assumed to be fixed throughout the building therefore every genome has an identical section 3.

10.14 Evolutionary operators

The same evolutionary operators described previously are applied to each rectangular partition. However, to ensure column continuity some additional steps are necessary:

- *Mutation*: Having selected which individual to mutate, the mutation operator randomly chooses, with uniform probability one partition of the building. It then selects an individual gene and generates a new variable for it. If the mutation operator selects a gene from sections 1 or 2 (Figure 10.1), the gene is replaced with a randomly generated value between 0 and the spatial limits of the rectangle (Figure 10.10). Unlike BGRID, which restricts the

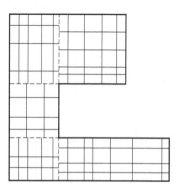

Figure 10.9 Invalid partitioning.

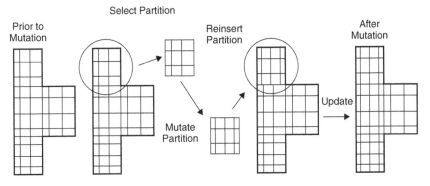

Figure 10.10 Mutation operator.

new spacing to a value between the two adjacent genes, the new system simply generates a random spacing and, if required, sorts the genome so that the column spacings increase from left to right. Having altered its genome, the section is placed back into the building and all adjacent sections are updated (Figure 10.10). This final step means the mutation operator is able to modify the building in only one location but the change ripples throughout the building preventing column alignments degenerating. This means that the mutation operator has significant impact on the form of the solution.

- *Crossover*: OBGRID employs a single point crossover operator (Goldberg, 1989), which exchanges part of the genomes associated with a section of the building. However once recombination has been accomplished, the altered sections are reinserted into the building and all other adjacent partitions updated (as with the mutation operator described above) as is shown in Figure 10.11.

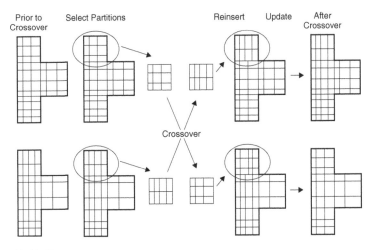

Figure 10.11 Crossover operator.

10.15 Illustrative example: an orthogonal building

The following test case is presented to demonstrate OBGRID's performance. The basic floor plan is given in Figure 10.12a and the partitioned floor plan is shown in Figure 10.12b. The parameters used are given in Table 10.2.

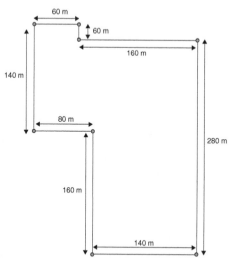

Figure 10.12a Orthogonal layout.

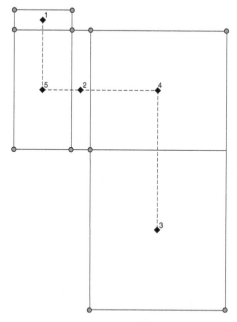

Figure 10.12b Orthogonal layout partitioned.

Table 10.2 Evolutionary algorithm (EA) tableau for orthogonal building

Objective	Evolve example building designs
EA	Genetic algorithm
Representation	String: real encoding.
Initialisation	Random initialisation (no seeding)
Raw fitness	Based on: overall height, column spacing compatibility and column spacing uniformity
Selection	Tournament (size = 2) with Elitism
Major parameters	P = 1, M = 100, G = 150
Evolutionary operators:	
Reproduction$_{prob}$	0.1
Mutation operator	Point
Mutation$_{prob}$	0.3
Crossover operator	One point crossover
Crossover$_{prob}$	0.6

10.16 Results

The genetic algorithm located the solution with the highest fitness in generation 97. The results for the column spacings are shown in Figure 10.13 with the other features being the use of a long span structural system with an average column spacing of 20m and mechanical ventilation. Other solutions for shorter span lengths were also generated by the system. With the fitness function used, these have lower fitnesses than the result shown but, as discussed above, the user may wish to look at these solutions and may also wish to amend the fitness function so as to better appreciate the sensitivity of the solutions obtained to the parameters used in the fitness function.

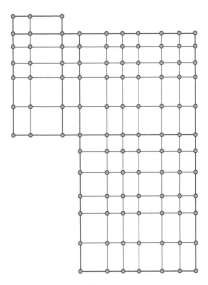

Figure 10.13 Best solution: generation 97.

10.17 Conclusion

This and the preceding chapter give an introduction to the use of genetic algorithms and, more generally, of search techniques for construction-related challenges. The work has been presented in the context of its use for design but it is equally applicable to many complex problems (e.g. Clarke *et al.*, 2009). Space restrictions have meant that only the basic techniques can be described. For those who wish to know more, the works referred to in these chapters give a good starting point.

11 Future technology for knowledge management

Rethinking knowledge technologies for our built environment; Map of ICT efforts and development in construction; Application-driven data integration; Application-driven semantic integration; Process-driven data integration; Process-driven semantic integration; Proposed semantic process-driven vision for the future of construction; Semantic process-driven vision in practice; Envisioning the future: Buildings as knowledge-intensive living systems; Conclusion.

11.1 Rethinking knowledge technologies for our built environment

In order to understand the need for future knowledge technologies in construction, it is important to take a broader perspective on our built environment and consider the challenges facing designers and stakeholders from the industry.

Recent events in the UK and world wide have highlighted the vulnerability of our buildings and stressed the effect of the environment, including the consequences of climate change in terms of weather extremes (high winds, prolonged heavy rainfall, flooding, drought and heat waves), rising sea levels, coastal erosion and water stress. From a societal perspective, the profiling of European (including UK) population demographics, coupled with behaviour and lifestyle changes, have resulted in diverse (including higher comfort) expectations which, if not managed, can hamper governments' sustainability targets.

Adaptation and resilience are the answer to the above challenges. Moreover, it is essential that we build the potential for adaptation and resilience into brief formulation, design, construction and facility maintenance methods. In this context, the industry needs knowledge technologies and processes that pave the way to sustainable and integrated building project development approaches with total lifecycle considerations.

Calls for research and development in the area of intelligent buildings and smart home environments have been made by several international and national government agencies and funding councils, the outcomes of which are reflected in several research and development projects and networks. However, the following can be noted in relation to the outcomes of the these initiatives: (a) lack of true multi-disciplinary approach and methodological interventions that factor in engineering sciences (advanced materials, mechanics, electronics),

Building sciences, architecture design approaches, resilience research (including fire engineering), sustainability sciences, and human sciences (comfort/behaviour); (b) Lack of total lifecycle thinking (from brief/concept design to facility dismantling and recycling); (c) Lack of value chain integration (designers, manufacturers, facility managers, utility services, etc.); (d) Strong emphasis on technology and lack of human-centric approaches; (e) monolithic or vendor/manufacturer specific closed solutions with lack of interoperability.

Any vision for the future of the built environment should address the following research questions:

- How do we confer optimum resilience, sustainability and continual fitness for purpose on our existing building stock?
- How do we deliver new or refurbished human-centric buildings that address lifetime requirements and that are capable of performing optimally within the constraints of unknown future scenarios?

The answer to the above questions requires a complete rethinking of the ICT equation for the industry and the role of knowledge. The chapter discusses and categorises construction IT research in terms of its underpinning technologies, based on two dimensions: semantic focus and integration approach; and then, attempts to define the future of knowledge technologies in the industry.

11.2 Map of ICT efforts and development in construction

Technology-led efforts described earlier in the book can be mapped according to a two-dimensional representation (Figure 11.1), developed with respect to two axes: semantic focus and integration approach:

- *Semantic dimension:* This dimension spans the whole spectrum of past, existing, and future applications with underlying semantics ranging from data structures conveyed through data models to rich-semantic representations through ontology. Furthermore, this axis conveys the following attributes: fixed versus flexible (In terms of function); context free versus context aware (with reference to the context in which information is viewed/used by individuals); one size fits all versus individually tailored/customised ICT systems. In essence, this dimension refers to research into IT support that is more specifically targeted and more easily composed/re-composed for individual circumstances using ideas from emerging information semantics, knowledge management and services research.
- *Integration approach:* This dimension represents the focus of research effort on a continuum from application and API centric, to process and people centric.

Examples of the research outputs are provided in each quadrant of Figure 11.1. The large arrows are used to indicate the relative effectiveness of the research

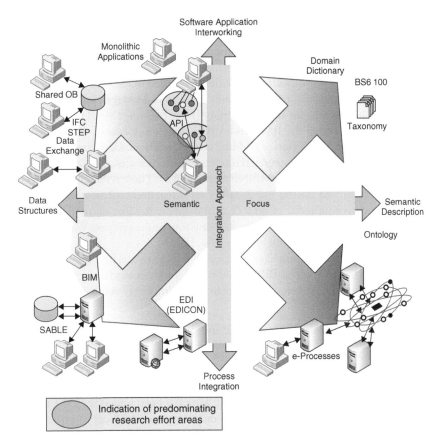

Figure 11.1 Map of IT research (adapted from Boddy *et al.*, 2007).

directions at the current time. Shading is used to depict potential for impact: light shading being less positive whilst dark shading is more positive

The following four sub-sections discuss research within each quadrant.

11.3 Application-driven data integration

Application and data integration has been a primary focus in construction IT research (Boddy *et al.*, 2007). Much of this work has for the last two decades been centred on the definition and exploitation of product data models for the construction industry. Numerous projects, both academic and industrial, have been undertaken in pursuit of this goal, some of which are mentioned here to give a flavour of the field.

Amongst the first efforts at integration were those born of the increasing use of CAD in design offices from the mid-1980s to present days. Here the necessity of transferring CAD data from one system to another resulted in de facto standards that persist to this day, such as the Drawing (or Data) Exchange Format (DXF) and the Initial Graphics Exchange Specification (IGES). More coordinated standards-defining efforts came in the form of the STEP application protocols for construction (ISO, 1994). This work, inspired by previous work primarily in aerospace and automotive fields, resulted in ISO 10303, part of the International Standard for the Exchange of Product Model Data. Latterly, the International Alliance for Interoperability (now, BuildingSMART: www.buildingsmart.com) defined the IFCs (Industry Foundation Classes), a set of model constructs for the description of building elements. The academic research community produced several integrated model definitions as described earlier in the book (including ATLAS, COMBINE and RATAS). These research efforts were generally predicated on the use of either an integrated tool *set* also furnished by the respective projects, or on a central database holding all model data for access by any application used in the construction project process via some form of adapter (Björk, 1989; Björk and Penttilä, 1998).

These research-led efforts have developed in recent years into industry initiatives around the concept of BIM (building information model) facilitated by a new generation of CAD environments from leading CAD vendors.

11.4 Application-driven semantic integration

Semantic resources are used as a means to enhanced integration of information and knowledge-based systems. The majority of semantic resources are constructed so as to be easily computer-interpretable for leveraging in other applications as appropriate. Further academic work has been undertaken to produce construction-oriented taxonomies and ontologies. The e-Cognos project, for example, generated a construction domain taxonomy and associated ontology based in part on several other classification and modelling schemes including BS 6100, UniClass and the IFC (El-Diraby *et al.*, 2005; Rezgui, 2007d).

Kalay (2001) introduced the P3 platform, a system of three interconnected databases each of which has a role to play in furnishing a semantically comprehensive and comprehensible picture of a project's data. Allied to these databases are Intelligent Design Assistants (IDeAs), which encode and store 'observer-specific information' (Kalay, 2001) and make interaction with the data in the databases possible. Halfawy and Froese (2005) propose Smart AEC objects as a step to better integration. Such objects feature, amongst other things, integral notions of the different views or perspectives held by individuals from different construction disciplines with respect to the object. Thus they incorporate at the model object level the kind of semantic data normally encoded into taxonomies etc. The proposal also describes how the IAI Industry Foundation Classes could be extended to implement Smart AEC Objects. The ISTforCE project (Katranuschkov *et al.*, 2003) uses an ontology-based access framework. Here, access to the product model data at the heart of the ISTforCE platform is provided through a mapping layer comprising the ontology. The ontology is in reality several different domain-specific ontology instances mapped to the same model data through a parent or top level ontology. It is argued that this furnishes a more user-oriented way of working with the model data in that concepts and operations common in the domain are used instead of those native to the model itself.

11.5 Process-driven data integration

Moving to the lower left quadrant, we see applications with more intuitive levels of abstraction. The SABLE project in particular has made progress in this respect having discipline specific interfaces to server-based IFC building models (SABLE, 2006). These interfaces including client briefing/space planning, architecture, HVAC design, cost/quantity takeoff and scheduling move closer to the process-oriented view of the project.

The field of 4D-CAD has attempted to integrate project scheduling with the building data model, primarily to better understand and coordinate the construction process. This integration obviously follows the project process and embeds that process into the software applications used in design. The integration task however, is typically a non-trivial one, taking considerable time and effort to manually create the relevant links between the project schedule and the various building model elements involved in any given stage (Koo and Fischer, 2000). This time and effort and the associated cost limits the utility of the 4D approach for smaller projects with commensurately smaller budgets or for evaluating multiple schedule options. Clearly a semi- or fully-automated integration of CAD and schedule in 4D would improve matters. Other moves to go beyond 4D CAD into nD, with tools for multi-dimensional interrogation and analysis of the building model have been proposed, such as the 3D to nD platform (Jongeling *et al.*, 2005; Lee *et al.*, 2002). Once again these systems work on a more task-driven basis to help project participants to better analyse and justify their design choices in an intuitive fashion.

Yet more academic projects have attempted to integrate varying degrees of process model into their overall project data model in order to link process to other project data e.g. (Christensen *et al.*, 1997; Jeng and Eastman, 1999). The process models proposed are actionable to differing degrees in support of this aim, thus incorporating a workflow into the project model and associated tools.

Also in this quadrant, there is a move toward distributed architectures offering services and service integration as part of the package. Many of the tools already mentioned incorporate elements of this, whilst others set out with distributed service integration and support for virtual enterprises among their primary goals. For example the OSMOS project (Rezgui, 2001, 2007a) offered extensible distributed services integrated with a central information management repository operated on an application service provider-like model. The ISTforCE (Katranuschkov *et al.*, 2001) project used similar principles. The DIVERCITY project (Aspin *et al.*, 2001) took a slightly different approach by incorporating services into a virtual reality collaborative design workspace, essentially a feature-rich fully immersive version of a virtual enterprise workspace.

It can be argued that the most process-oriented technology in the data/process quadrant is EDI. EDI continues to be used widely in a broader industrial/commercial context in the form of the EDIFACT standards (ISO, 1988). EDI evolved into EDIFACT over many years from the 1960s onwards and numerous sub-dialects have been defined for specific user communities, EDICON being that for construction. Research shows however that the actual use of EDICON in the construction industry could best be described as sparse and getting sparser as years go by (Akintoye and McKellar, 1997; O'Brien and Al-Soufi, 1994).

11.6 Process-driven semantic integration

Construction domain-specific work in this area is less common than in most others, though some researchers have begun to look this way. Fellows and Liu (2003), for example, have proposed a distributed ubiquitous computing environment based on multiple collaborating services presented to client devices through what they term 'active mediators'. They also propose an element of semantic service description using ontologies, and service choreography based on Process Specification Language (Schlenoff *et al.*, 1999).

The Inteligrid project (Turk *et al.*, 2004) envisions computer-integrated construction over a grid architecture where services are offered and consumed in a dynamic environment that is opaque to the end-user. These services to some extent map onto processes in the construction lifecycle, though the vision seems to indicate that services would be mainly of a technical nature such as thermal load calculation or structural analysis rather than services as processes or workflow in a business sense. In common with other projects, ontologies are seen as a crucial element in mediating between various heterogeneous service implementations and also between services and users. This is very much ongoing research at the time of writing, with implementation details still emerging from the project.

Aziz *et al.* (2006) espouse a vision based on web services, again dynamically combined as and when needed. The primary focus of their exposition is that of extending the computing environment beyond the office with mobile devices for construction site workers, supply chain operatives etc. The task of service discovery and composition/choreography is delegated to autonomous agents, acting on behalf of system users to furnish their computing needs in a multi-agent system. The granularity and nature of the services envisioned is not specified, making it hard to say if they are process- or data-oriented, though once again ontologies are seen as the glue allowing for their seamless integration.

Grilo *et al.* (2005) propose a system of process-oriented services. Here the idea is to use the Model Driven Architecture (MDA) (Soley, 2000) paradigm from the Object Management Group at two levels. Firstly it is used to model application semantics at the meta-level for the purpose of general interoperability, and secondly it is used at the industry sector level in order to define sector-specific services. These services could be modelled such that they are analogous to processes within a construction context. The proposal also suggests the use of the currently popular and evolving service-oriented Architecture at the application level to confer additional service discovery and interoperability benefits and as a platform-independent representation of the application interactions. It is suggested that by using such an approach, services based on heterogeneous platforms could be integrated into a single process-oriented computing platform supporting a particular part of the construction project process. This approach, then, relies heavily on widely agreed and adopted vertical domain standards (the sector-specific models) for its operation and success.

Despite the lack of specific attention in this area, we would assert that process integration through semantically described and mediated composite services offers great potential for supporting the crucial communication channels between actors, and ultimately improving the end-user's experience of IT in construction.

11.7 Proposed semantic process-driven vision for the future of construction

Data integration efforts lack an overall context and referential as these efforts are driven by the need to make software applications interoperable. Whilst very structured data, such as a building model, has a well-defined internal context, its place in the project is less certain in terms of its relationship to other information and the people who use it. An integration that places information in a context that is easily understood by those working on the project, i.e. one abstracted away from file stores and computers and geared more towards processes and project context, provides a superior arrangement. Both STEP and IFC could be argued to incorporate some aspects of context, modelling as they do some project scheduling and organisational elements of the domain. This however still fails to incorporate information that exists outside of the model and is not executable as a process in its own right.

Also, for large international standards efforts, agility is something of a problem. Once the standard is agreed, changing it can take a considerable amount of time, which in an age of rapidly evolving business needs can turn a formerly helpful system into a hindrance.

Given the problems outlined with data-level integration, an environment based on business and project processes presents more potential in terms of modelling the real project environment, i.e. the natural level at which people interact with their work in an organisational or project context. There is already a trend towards business process integration in the wider commercial context, with many vendors offering 'solutions' in the domain (Boddy *et al.*, 2007). These systems tend normally to be focused on integrating legacy applications into a process workflow in a very organisationally introspective way. In an environment typified by short-term partnering arrangements between multiple organisations, a more eclectic approach to integration is required to support the

creation of short-lived project-specific business systems integrated and operated at the process level. Here we draw on object-oriented principles in that
we envisage a number of components, each furnishing a small piece of functionality in a fully encapsulated independent way. These components would
be published as web services and composed into higher-level business process
components, again self-contained (Rezgui, 2007c). Each business process would
service some pertinent part of the construction project lifecycle. It may be the
case that the same process component is published in several different guises to
suit different organisational or disciplinary perspectives on the project process
it supports. For example the view of a component supporting commissioning of
the HVAC installation from the engineer's point of view would be different to
the view required by the client who ultimately needs to approve it. The components therefore will also be required to adapt to the context from which they
are executed. This model of arbitrary combinations of process components, or
e-processes as one might call them, allows for greater flexibility in the definition
and construction of business systems to support construction projects. Using
this model it should be possible to select best-of-breed process components on a
project-by-project basis, the systems effectively configuring themselves based on
a description of process needs. Even reconfiguring the project business system
during the life of the project to reflect changes to the description of needs would
be feasible.

The question of information in context becomes less of an issue with process-
based systems as information is firmly rooted as the fuel by and for the process
(Boddy *et al.*, 2007). A clear and thorough analysis of information needs at each
stage of a project process will serve to both identify data needs and contextualise
that data within the framework of the process. This leaves scope for continuing research into information structuring and management, but primarily as a
means to better understand and situate it in the process than as an end in its
own right as has often been the case to date. Further, such research could move
to a more knowledge-based mode where a focus on integrating unstructured
data through semantic analysis or indexing, perhaps employing techniques from
the information extraction field, would help to locate data automatically in the
frame of the process. Benefits may also be derived from uncovering previously
unseen linkages between various elements of project data using such analysis
methods.

This vision does not necessarily mandate that the whole project process be
mapped and serviced by process components from the outset. It is an integral
part of the vision that as much or as little of the project process as is feasible or
desirable should be supported. Thus it is possible to integrate processes incrementally as the underlying functionality becomes available and is proven to be
effective by the early adopters.

The technology to do most of this already exists or is being developed. For
example, in order to have process components (essentially a composite web service) integrate seamlessly as we have envisioned, it will be necessary to describe
them at a semantic level in addition to the basic input and output messages

defined by the standard WSDL service description. For this purpose Ontology Web Language for Web Services (OWL-S (OWL-S Coalition, 2005)) could be employed. OWL-S allows for the ontologically grounded description of web services, both atomic and composite, making the description computer interpretable. The language supports the notion of profiles that describe both what the service (or process component) offers and the requirements of a searching agent that are met in the service, with each service potentially having multiple profiles. Thus searching for a service that meets the requirements of part of a model of process needs can be automated. Further work is underway to design a method for describing and reasoning about the dynamic behaviour of a service prior to its invocation to further ensure compatibility between services (Solanki *et al.*, 2004). OWL-S also allows for other ontologies, taxonomies etc. to be used to describe various sub-parts of the service, for example in order to classify the service into a construction-specific scheme for our purposes. It is here that an element of discord may be re-introduced by the use of disparate ontologies in the underlying service descriptions. There is however a considerable ongoing research effort in the area of inter-ontology mapping and ontology merging (Noy *et al.*, 2000), which would allow for multiple ontologies in service descriptions to be leveraged in a consistent manner. It is also at this sub level that we would anticipate IFC, given its taxonomy oriented nature and substantial industry backing, to play a pivotal role in providing the ontological backbone in support of e-processes. A number of efforts have been initiated in this direction as reported in Barresi *et al.* (2005). The Web Service Modelling Ontology (W3C, 2001) proposal from the W3C has constructs similar to OWL-S but includes the notion of mediators. Mediators are specific elements of the model that address heterogeneities between the various other parts of the model such as multiple ontologies used in the description of services.

11.8 Semantic process-driven vision in practice

There is a need for organisations to migrate their legacy systems to higher order applications capable of engaging in dynamic modes of collaboration to support distributed business processes. This requires a change of focus from intra-enterprise system integration through agreed data structures to inter-enterprise business process integration through smart composition of web-serviced applications.

Service-oriented computing is becoming the prominent paradigm for leveraging inter and intra enterprise information systems, creating opportunities for smart organisations to provide value added services and products. The benefits of web services include the decoupling of service interfaces from implementation and platform considerations, the support for dynamic service binding, and an increase in cross-language and cross-platform inter-operability (Ferris and Farrell, 2003). This form of computing is now moving from its initial 'Describe, Publish, Interact' capability to support dynamic composition of services into reinvented assemblies, in ways that previously could not be predicted in advance (Curbera, 2002; Rezgui and Nefti-Meziani, 2007).

Semantic e-processes are typically designed, developed and deployed by enterprises that want to compose internal capabilities with third-party capabilities, either for internal use or to expose them as (value-added) e-services to customers. They can be described as the smart aggregation of services that are captured in the form of a Composite service complying with well-defined business rules. They have the ability to be re-used outside their scope and become generic services (Rezgui, 2007c).

It has been argued earlier in the book that a construction project should be operated as a construction virtual enterprise. This would involve facilitating and supporting the execution of distributed business processes, implemented through a coordinated composition and invocation of web-enabled, service-oriented, corporate enterprise information systems. Shared workspace environments provide an interesting paradigm to implement construction virtual enterprises. A shared workspace refers to an online web environment involving a cohesive community of authorised actors, united for a business or practice purpose (Rezgui, 2007c). The concept of 'actor' refers to a person or an organisation performing one or several tasks or activities, through an assigned role, within the context of a construction project. Access to data/information/knowledge from within a shared workspace is organised via dedicated elementary or composite services, invoked by actors through their assigned role(s). Services are geared towards supporting and implementing business processes. Shared workspaces can be defined at different levels of granularity; these can range from supporting collaboration within a complex construction project to nurturing a small community of practice. Implementing service-oriented shared workspace environments involves three key generic 'roles' defined at the service infrastructure level (illustrated as layers in Figure 11.3):

- *Virtual Enterprise service provider (VESP)*: This has the responsibility of managing the entire service infrastructure (e.g. servers, computer resources, services, etc) and allocating shared workspace environments to potential construction clients to host their projects. This involves hosting the VE core infrastructure through provision of and access to both core services and third party services (TPS). Core services refer to services necessary for the basic operation and management of services, including TPS that have the particularity of being provided by third party entities, namely, third party service providers. VESPs, through the core services, have the capability to host multiple VE projects and to make available different services (both core and TPS) to various VE projects. VESP can also play the role of application integrators, providing help and assistance to migrate legacy applications and legacy systems to Web services.
- *Third party service providers (TPSP)*: These represent various companies, including software houses, interested in making their software application(s) accessible through a service-based middleware solution hosted and managed by the VESP. These companies have their services published within the VESP UDDI (Curbera, 2002) registry. Typically, these services would

be geared to serving a particular purpose for the VE to which they are being made available. Examples of services offered by TPSP include structural dimensioning service, HVAC simulation service, procurement service, and facility management (FM) service. TPS are offered by organisations that procure the service implementation, supply the description of the service, and provide related technical and business support. These can also represent service aggregators that consolidate multiple services into a new, single service offering.

- *Virtual Enterprise clients (VEC)*: The VE through its allocated shared workspace will involve stakeholders representing construction companies collaborating within the context of a project. This collaboration is supported and enabled through the VESP platform. While one company, acting in the capacity of VE manager, would configure and administer the VE, others would make use of the core and TPS services made available to the VE project.

While the above generic roles represent the underlying concepts of a potential service infrastructure for the construction industry, business roles (such as architectural design, structural design and quantity surveying) are defined at the level of the workspace (implementing a VE) and are directly assigned to Virtual Enterprise clients. The latter interact directly via their business role with services allocated to the workspace in which they operate in accordance with their assigned service invocation rights (Rezgui, 2007d).

A construction project (assimilated to a virtual enterprise) has a lifecycle (as illustrated in Figure 11.2) at the end of which the project is ended, the infrastructure dismantled, and the information and knowledge generated throughout the lifecycle archived and re-used for various value-adding purposes.

In practice, the company acting as the VE service provider will initially agree a contract with that client depending on the client's specific VE requirements. A contractual agreement is made, and a dedicated VE project workspace is set-up. This involves the VE service provider allocating the necessary server(s), computer resources, logging facilities, required core and third party services. Within the activity, a VE project administrator will be registered and the VE project will be initialised with the services selected for its operation, and configured based on the requirements identified under the terms of the contractual agreement. The VE configuration involves:

- Configuring the assigned services to meet the needs and project requirements of the VE;
- Identifying actors to be involved in the specified VE project;
- Defining and delineating the roles within the specific VE project;
- Assigning the defined roles to the actors;
- Launching the VE project in accordance with the configured project services, agreed VE procedures and assigned roles, under the control of the project management committee.

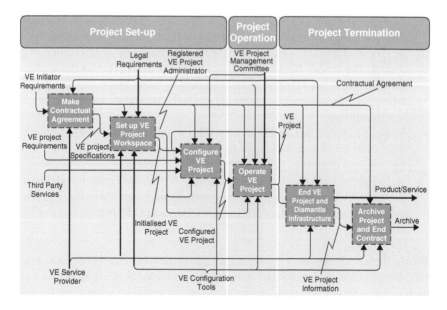

Figure 11.2 Provision and maintenance of virtual enterprise projects.

Once configured, the VE project can enter into an operation stage. The operation of a VE project is done under the control of a project management committee and is performed by the VE project administrator. This involves:

- Managing and administering all actors participating in the current project;
- Managing roles as assigned to respective actors;
- Managing access rights to support the information needs of the actors participating in the VE project (e.g. sharing, exchange, communication, dissemination, archiving, workflow and scheduling). Access rights are attached to roles in the VE project;
- Using services to perform business processes. (This activity may also include the need for training in the use of unfamiliar software applications);
- Managing VE project services. This comprises of all the activities the VE project administrator needs to be able to perform in order to ensure the assigned services for the specific VE project are available to the actors of that project, thus enabling continuation of the operational VE project.

Once the project ends and the objectives of the VE project are fulfilled, the VE is ended and its associated software infrastructure dismantled. This activity occurs at the end of the VE project's lifecycle and represents the actions required to complete dissolution of the current VE project. The final stage involves archiving all of the information produced during the VE project lifecycle according to the contractual agreement. The services that underpin the

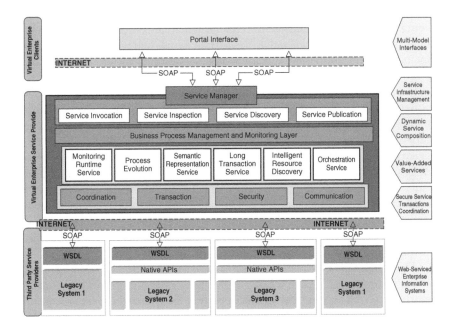

Figure 11.3 Proposed architecture for process-driven integration.

above construction VE service platform are articulated around an ontology. The ontology is used to elicit the semantics of each service in terms of data structures and enable service composition and e-process support.

There is a need for adapted and user-friendly environments to define, monitor and fine-tune the execution of (long-lasting) e-processes. These environments can be used to define a project process and to map that process into a series of requirements specifications for dispatch to the web services composition framework (OWL-S/WSMO etc.). The matching agents in the framework would then be able to assist in the selection of suitable service offerings, or in the longer term make those selections autonomously to furnish a software suite tailored to the project process initially expressed (Boddy *et al.*, 2007; Rezgui and Nefti-Meziani, 2007).

11.9 Envisioning the future: buildings as knowledge-intensive living systems

As argued earlier in the chapter, we need to rethink construction in the light of the societal and environmental challenges faced by our built environment. Traditionally, buildings, and their constituent materials and components, are designed and manufactured to meet a prescribed specification; material degradation is viewed as inevitable and mitigation necessitates expensive maintenance regimes. More recently, based on a better knowledge of biological systems, materials that have the ability to adapt and respond to their environment are being developed. This fundamental change in material design philosophy facilitates the creation of a wide range of 'smart' materials and intelligent structures, including both autogenous and autonomic self-healing materials, and adaptable, self-repairing structures that can positively react to their environment.

It is essential to embed adaptability and resilience within materials and components, thereby enabling the capability to adjust to any potential state change induced by environmental stimuli. These smart materials and components should have the capability to collaboratively respond to these stimuli/adverse events through a 'modulated response'. This response is embedded within the design and manufacturing process and the 'modulation' is achieved through embedded sensing capabilities that continually establish the state of the building and the role of the smart component in the building integrity. Thus the state of the building can be defined at the moment of the stimuli and adaptation (i.e. building performance) and emergency response can be optimised on the basis of this information.

This suggests that we need not only to have a good understanding and model of our environment, a comprehensive model of our buildings, but more importantly better understand the interactions that take place between occupants with their buildings and the impact this has with our environment (Figure 11.4). Traditionally, building conceptualisations have been captured through static models as noted earlier in the book, i.e. IFCs or BIM. We need more comprehensive building ontologies that have the potential to provide real-time accounts of a building state and performance.

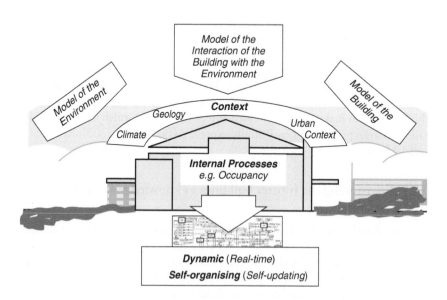

Figure 11.4 The building in its environment.

This suggests a number of knowledge related challenges for the construction research community, including:

- *Smart materials*: There is a need to rethink construction materials and exploit recent development in material sciences to promote the commercialisation of a new generation of materials that provide condition feedback and embed shape memory, energy sensitivity and self-healing properties (enabled by self-triggering biological or other processes).
- *Smart components*: Based on the emergence of the smart materials concept, there is a need to develop a new generation of smart (parametric, modular, inter-operable and context-aware) building components with embedded intelligence that enable buildings to be assembled from, or refurbished/retrofitted with smarter engineered products and systems, while providing greater flexibility to satisfy larger customer choices and changing regulatory requirements.
- *Building ontology*: There is a need to enhance current IFCs and BIM efforts and develop a dynamic and total lifecycle model of a building that provides forward and backward compatibility and regulatory compliance checking, real-time performance accounts, simulates future adaptability requirements, ensures optimum resiliency and fitness for purpose.
- *Service-based software delivery*: The construction industry needs more robust and scalable software/hardware infrastructures, using a cloud computing model, that will host a wide range of design and manufacturing services providing business process orchestration and workflow integration across the project lifecycle and value chain.
- *Smart alliances*: The construction industry needs new organisational forms and contracting methods to deliver the above challenges and mobilise the right distributed skills, competences and resources to achieve economies of scale in the production of standardised processes and product development approaches, with economies of scope in various stages of assembly and maintenance of a building or facility.

This necessitates methodological interventions and approaches that factor in multi-disciplinary and multi-faceted perspectives on common built-environment issues. In that respect, a systems thinking perspective is essential as it provides a foundation for adaptable buildings. A systems philosophy demands that an uncoordinated approach is replaced by a framework in which the identities of the separate parts are subsumed by the identity of the total system. In engineering terms, using a systems engineering approach, the individual elements and subsystems of a building are designed and fitted together to achieve adaptability and resilience. Furthermore, the construction IT community needs to adopt a systems dynamics approach to building adaptability to map the dynamic forces that affect building performance and aid understanding of how the different variables and subsystems within an adaptable building system interact. As illustrated in the figure, a holistic approach would need to factor in manufacturing

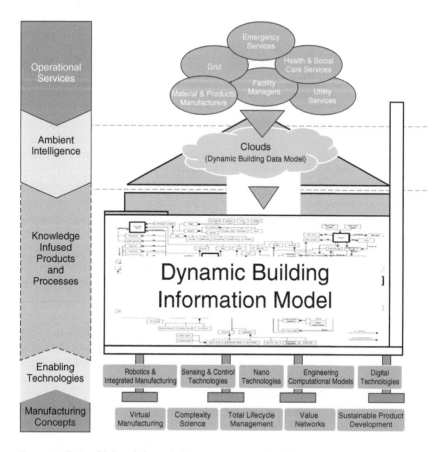

Figure 11.5 Total lifecycle knowledge perspective on buildings.

concepts and enabling technologies into ambient intelligence in the form of clouds of data that provide means for total lifecycle operational services.

11.10 Conclusion

This chapter has provided insights into the future development of knowledge systems and more generally IT in construction. Two major developments are expected to enhance ways in which we design, construct and operate buildings:

(a) IT will expand from supporting basic and discrete IT functions to supporting long-lasting processes that span the total lifecycle of the project facilitated by true building ontologies.

(b) These ontologies will have a strong dynamic dimension facilitated by smart materials and building components that operate within managed systems and provide means of integrating stakeholders across the supply chain and total project lifecycle.

Knowledge technologies will take various forms, ranging from knowledge and competency management to advanced decision support systems. However, these technological advances need a more technology-mature knowledge work force that uses adapted, including virtual, modes of working and conducting business. This forms the focus of the next chapter.

12 Knowledge-infused alliances of companies

Virtual teams in construction; Partnering in construction; Construction alliances: the case of SMEs; What is an alliance? Knowledge needs for transient SME alliances; Business opportunities management in an alliance; How can an alliance be sustained? Role for information and communication technologies; Promoting innovation in an alliance; Conclusion.

12.1 Virtual teams in construction

Chapters 2 and 3 discussed the importance of virtual teams in the construction sector, and the way these collaborate on projects to deliver the products and processes necessary to design, construct and maintain a building or facility. The characteristics of the construction sector suggest that in researching, developing and evaluating potential forms and solutions for virtual teamwork, the human and organisational aspects require close attention. This means that social and, ultimately, economical considerations have to be made rather than concentrating on technology alone as has been traditionally the case in this sector.

A construction project, it has been shown, can be assimilated to a virtual enterprise (VE) that spans the entire lifecycle of a project. Virtual teams rely on cooperation between dispersed stakeholders working towards a common purpose. In order to efficiently meet this purpose, the goals of the virtual enterprise (i.e. project) need to be shared and embraced collectively.

A construction VE can be defined as a grouping of virtual teams (representing various disciplines) bound by contractual agreements that collaborate at different timeframes of the project lifecycle. These virtual teams require adapted controls so that to increase their effectiveness at an individual, team and project level (Rezgui, 2007a). The contractual agreements should include both the extent to which information and knowledge are managed and shared and the degree of control employed.

In this context, the service-hosting concept presented in the previous section (11.8) provides the technological infrastructure to:

(a) Support the central business processes;
(b) Allow integration of systems, interoperability between disparate applications, and management of long-lasting processes; and,
(c) Support the management of interactions between individuals and teams.

However, existing research has identified important socio-organisational issues inherent to the project-based nature of construction that must be addressed and blended successfully toward the shared VE purpose. These include issues related to team identification and trust, and more generally to the challenge of virtual project management (Rezgui, 2007a). Moreover, the migration path to successful virtual team working is grounded in social and organisational elements that engage all stakeholders in a manner that promotes and engenders trust. This involves an exercise in change, which requires new mechanisms to enable participation and communication within and across organisations, and more importantly on projects (Rezgui *et al.*, 2005).

The issue of team identification impacts on the cohesiveness of a team, including the ties that bind teams (Scott and Fontenot, 1999). Construction team members often are involved in more than one project or team at a time, making it difficult for them to define their many identities (Rezgui, 2007a). In fact, identity boundaries are present when some members of a team are not fully dedicated to the team, either because they are working on multiple projects with multiple teams or because their teams are nested within larger teams. This can have powerful effects on behaviour (including, satisfaction, trust, and performance) as highlighted in related literature (Espinosa *et al.*, 2003).

In this context, the project coordinator or manager is a strong target of identification (Rezgui, 2007a). Team leadership in distributed settings is therefore critical to team effectiveness (Connaughton and Daly, 2004). Leading a virtual team not only involves the communication complexities, but also requires a certain shift in the project leadership approach. Virtual project managers should exhibit a number of essential attributes, including: leadership, results catalyst, facilitator, barrier burster, business analyszer, coach and living example (Fisher and Fisher, 2001).

Three key aspects emerge from existing literature and can be recognised as ingredients for successful migration from traditional organisational forms to virtual team working within the context of a construction VE (Rezgui, 2007a):

- *Knowledge sharing*: existing literature in the field confirms that facilitating VE members' access to shared, well-structured and unambiguous information and knowledge improves communication and cohesion amongst the members of the VE.
- *Trust and cohesion*: trust among leaders and team members may be swift yet fragile (Jarvenpaa and Leidner, 1999), and members' identification with the team and leader may be challenged over distance (Connaughton and Daly, 2004). In fact, identification is the psychological tie that holds virtual team members together, and is therefore important for achieving desired

outcomes (Connaughton and Daly, 2004; Wiesenfeld *et al.*, 1999). Trust is positively related to team members' identification with their leader, in both distanced and proximate relationships. Equally, open communication channels and participation and involvement in decision-making enhance sharing of information and facilitate virtual teams cohesion, which in turn promotes trust. Indeed people work together because they trust one another and special attention to building trust should be given throughout the VE lifecycle. The following recommendations can be formulated (Rezgui, 2007a):

1   Include face-to face interactions when possible during the virtual team lifecycle and in particular during the inception stage where the vision, mission, and goals can be communicated and shared.
2   Give equal access to information, including project status and progression.
3   Involve project coordinators experienced in virtual team management.
4   Develop strong communication and collaboration protocols, including code of conduct, standards for availability and acknowledgement
5   Appoint an experienced project (i.e. virtual enterprise) coordinator.
6   Select team members with aptitudes to work in a virtual setting.

• *Continuous learning*: As will be discussed in the following chapter, enabling a VE requires not only the implementation of innovative technologies but also new working practices, and organisational structures and cultures. The potential benefits of such innovation can only be realised through individuals at all levels learning and developing a considerable array of new capabilities. To this end it is recommended that the organisations wishing to deploy a VE consider the extent to which that training should be planned and implemented, and prepare a programme/documentation for the purpose.

12.2 Partnering in construction

While team working is what characterises construction projects, adversarial relationships between firms involved in teams have led both to poor quality and productivity in projects. This is due to a resistance to information and knowledge sharing, and to major barriers to learning and capitalising on lessons learnt from past experiences (Franco *et al.*, 2004). Partnering arrangements that transcend the duration of a construction project are essential for a healthy industry (Latham, 1994). This engenders trust (Bennett and Jayes, 1995), promotes knowledge sharing, long-term business collaboration, and investment in technology. This is emphasised in the Egan report (Egan, 1998) which identified partnering as a mean for gradually migrating from lowest price to performance driven contracting.

Large client organisations (such as the National Health Service in the UK) have played a key role in promoting partnering on projects. This was motivated by (a) the size of their building stock and complexity in managing/operating their buildings, (b) the need for total lifecycle considerations, (c) the need for long-term relationships and stability of arrangements between parties involved,

(d) the overall desire to reduce costs (including tendering costs), and (e) the easing of a claims-based mentality (Franco *et al.*, 2004).

Partnering arrangements involve clients, contractors, construction engineering firms and a large proportion of small and medium sized enterprises. In fact, SMEs value partnering as this provides them with a guarantee for business continuity (Rezgui and Miles, 2010). More generally, partnering is perceived as playing a key role in the generation of feedback and promoting learning processes, which in turn have been identified as critical missing processes in conventional construction arrangements (Bennett and Jayes, 1998; Franco *et al.*, 2004). However, construction partnerships, like strategic alliances in general, are less natural learning entities than are single organisations (Larsson *et al.*, 1998). Argyris (1999) points out that the development of trust is crucial to successful learning. Lack of trust can result from a number of situations, including past collaborative experiences, or clients driven by their own agenda.

In essence, any proposed alliance-based business model must factor in a number of issues, including the ones summarised in Table 12.1.

Table 12.1 Issues to be tackled for a holistic understanding of construction alliances

Issue	*Typical current practice*	*Challenges for partnering*
Financial risk	The main/lead contractor bears the risk as *the* contractor (whether building lead or design lead). A fixed price is agreed and maximum responsibility is passed on to subcontractors.	In a collaborative team from within an alliance, how will income, financial risk and responsibility be shared?
Motivation	Each sub-contractor is independently sub-contracted and his main concern is his bottom line profit from the job.	How does belonging to an alliance affect attitudes? Will alliance partners better support each other collaboratively in the event of problems?
Discipline	Contracts lay down the rules that discipline relationships and courts of law rule on disputes.	What codes of behaviour should there be in an alliance? What authority/ arbitration should there be?
Trust	Where participants may work together again perhaps for the same lead contractor or client, then good reputation is important. Cheapest price sometimes comes with least reputation.	How should performance of alliance members be measured and fed back? How open should feedback be to the whole alliance family? How are poorly performing members dealt with?
Liability Insurance	Insurance covering the long-term risk of a built facility from faulty design or workmanship is high. Lead contractors doing many projects spread the risk and gain insurance discounts.	How should insurance be handled in an alliance? How can a successful track record be measured when there may be different mixes of alliance members in different projects?

Hence, successful partnering or alliance formation involves: (a) an understanding of the nature of the inter- and intra-company collaboration and teamwork that take place on construction projects; (b) understanding of current barriers and limitations to collaboration and network formation, and (c) appreciation of the factors that facilitate the proposed new forms of business model adoption and use in various situations and work settings.

12.3 Construction alliances: the case of SMEs

Given the dominant and fragile nature of SMEs in the construction industry, there is a need as well as an opportunity for SME alliances to be established, forming virtual networked organisations that can compete in better conditions than individual SMEs.

An alliance will bring together SMEs in a relationship of trust to provide holistic competence in an area (e.g., energy, sustainability and integration of resources). The alliance has the potential to market this capability and achieve a recognised industry 'branding' much as a 'big name' player. Moreover, because the alliance will work together frequently (though not exclusively), it can standardise processes and information flows, develop trust, dispute resolution strategies and shared business models (Rezgui and Miles, 2010).

Thus, the *vision* is a transformation of production business models for construction SMEs, to achieve economies of scale in the production of standardised processes and products with economies of effort whilst providing flexibility to satisfy customer choices (Rezgui and Miles, 2010). The adoption of alliance modes of operation will promote business process innovation and allow SMEs to compete in new ways, get better reward for their work, gain greater financial strength, which in turn will give them the financial capability to move forward and develop their products.

The automotive industry, considered to be a beacon for industry, has long since incorporated the role of the 1st tier supplier, i.e. a company capable of design, procurement, assembly and delivery under JIT (Just in Time) methods of a whole assembly. This has an impact on the supply/value chain by forcing the emergence of knowledge-based alliances. Construction SMEs should learn from this by forming cooperative production networks to:

- Link their core activities to those of other SMEs to achieve an 'integration of competencies and technologies', so expanding competitive business opportunities, and their potential for joint innovation in integrated solutions;
- Rely on virtual business models and operations, enabled by SME friendly ICTs, which can facilitate the implementation and diffusion of sustainable design ideas and promote new ways of production and integration in construction.

While partnering is a common form of contracting that is transaction-driven and involves two or more contractors, an alliance has a more strategic and formal dimension, involves a supply chain, responds to niche opportunities, and relies essentially on information and communication technologies for its business operations and long-term sustainability (Rezgui and Miles, 2010).

12.4 What is an alliance?

A number of characteristics of 'being an alliance' or 'being in alliance' have been noted (Rezgui and Miles, 2010) and synthesised below:

- It is a state that is stable; it is a long-term 'marriage' of parties.
- Partners in an alliance 'know each other' (or are in the process of actively achieving close relationships) and are in a state of preparedness to act together when the need arises.
- Even when not actively engaged in a common project, there will be 'team building' and 'team training' that promotes alliance capability and preparedness.
- It is a union or association of independent entities operating peer to peer that respects complementary roles (as in a family). They have freely 'contracted' to work together and help one another within the envelope of the alliance.
- There is a high level of shared interest in the common good as well as individual benefit, since there may be partners in the alliance that would ordinarily be regarded as competitors.
- Trust and transparency are the 'critical success factors' that enable competitive individuals to 'live together' in a relationship of 'give and take' which prospers the alliance and rewards the partners fairly.
- Disputes are resolved in concert, quite possibly involving alliance members that are not directly involved in the dispute. What the alliance community feels fair and right is the disciplinary peer pressure that is brought to bear.
- Conversely, there is mentoring that takes place to raise individual skills and competence, to create a fitter whole.
- The alliance is promoted to customers more than individual partners, although individual circumstances may dictate the profile of a particular partner being highlighted.
- The alliance name should become a branding and a 'guarantee' of quality. The alliance becomes the entry point for customers knowing that a 'tailored team of fit for purpose partners' will be marshalled rapidly to evaluate requirements.
- In principle there should be little that prevents a partner being in more than one alliance provided that it is known and provided they are not directly competing alliances. This would work well in situations where there are narrow focuses in, for instance, niche areas.
- Alliance partners are scattered geographically, but on the virtual plain they are in exceptionally close proximity that may be thought of as in the 'room next door'. Partners do work and produce things (think of them as prod-

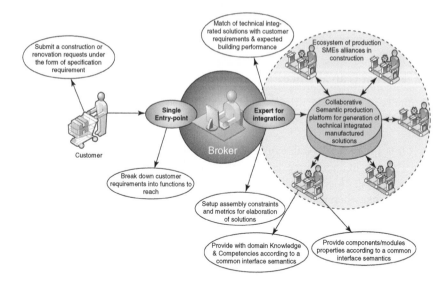

Figure 12.1 An alliance in practice (Rezgui and Miles, 2010).

ucts of work) that need to come together – strategies, requirements, plans, designs, components, assemblies, finance etc. A whole bundle of physical products and tangible services that need to be coordinated in time, space and budget through what is essentially a distributed production plant.

Finally, an alliance of independent peer organisations requires a 'management' hub of some kind (as illustrated in Figure 12.1). This would ensure (a) coordination of alliance strategic direction, (b) marketing the alliance, (c) handling contractual and advising on legal issues, (d) negotiating common resources (e.g. insurance and equipment), and (e) managing business opportunities. The latter is initially a bridgehead to customers who enquire but it becomes the communication channel between the alliance and the appropriate partners for the particular opportunity, including their selection from the cluster of partners in the alliance. It is perceived as a brokering role.

12.5 Knowledge needs for transient SME alliances

Based on the above, a number of technical and process requirements have been identified to support knowledge-infused SME alliances (Rezgui and Miles, 2010). Three categories of knowledge-oriented functionality emerge (Figure 12.2), namely:

Functionality to support alliance operations management:

• *Alliance formation and membership management*: SME member identification, registration, profile definition and membership validation/acceptance and management.

- *Alliance nurturing and long-term management*: functionality to support social networking within the alliance, promoting trust and shared responsibility management, supporting negotiation and conflict resolution, and functionality to support financial incentives and reward management.
- *Alliance creation and operations management*: functionality for roles and responsibilities such as participation in decision-making, SME safeguards regarding information and knowledge, resource scheduling in an alliance, and arrangements for cost-effective software use in SME alliances.

Functionality to support alliance business management:

- *Customer relationship management*: integrated and socially-oriented approach to managing customers.
- *Business opportunity management*: functionality to exploit business niches in the surrounding local, regional, and international business environment, and identify/nurture business opportunities.
- *Alliance branding and marketing*: functionality to sustain a vibrant and positive (with a focus on sustainability and social corporate responsibility issues) outfacing image of the alliance.
- *broker/customer management*: functionality supporting the SME broker in channels of communication inside the alliance virtual factory and outside to customers, regulators, suppliers etc.

Functionality to support alliance capability, learning and innovation management:

- Functionality to support and nurture knowledge sharing and creation, including best practice codification.
- Functionality to support awareness raising through push and/or pull mechanisms (dissemination/advertisement/broadcasting of SME information days, brochures, newsletters, etc.).
- Innovation management environments including legal, IPR, contractual and cultural diversity issues.
- Functionality to identify SME learning gaps and training needs.
- Functionality for self-learning SME members, including provision of online learning/training facilities.

The proposed functionality should be delivered using a service-oriented architecture. This is based on a set of loosely coupled components available as services, delivered using an application service provider (ASP) like model, as discussed in section 11.8.

A number of key issues emerge from the literature as essential to support effectively web services, including:

- Coordination (to manage interaction between services and coordination of sequences of operations, to ensure correctness and consistency).

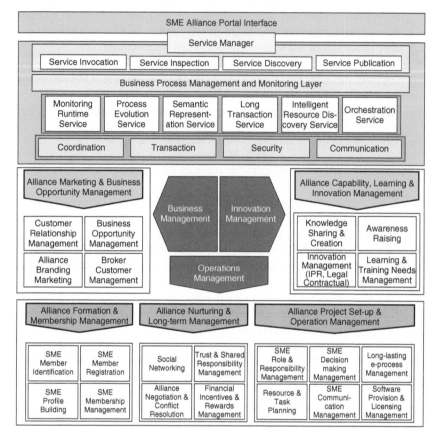

Figure 12.2 The SME alliance portal architecture.

- Transaction (to manage short-duration/atomic and long-running business activities).
- Context (to adjust execution and output to provide the client with a customised and personalised behaviour: may contain information such as a consumer's name, address, and current location, the type of client device, including hardware and software that the consumer is using, or preferences regarding the communication).
- Conversation modelling (to facilitate service discovery and dynamic binding, service composition model validation, service composition skeleton generation, analysis of compositions and conversations and conversation model generation).
- Execution monitoring (involves either centralised or distributed execution of composite web services).

Solutions to the above issues exist with varying levels of sophistication and complexity. These certainly form research avenues currently addressed by the web services community.

12.6 Business opportunities management in an alliance

The proposed model for an alliance involves a flat and flexible management structure articulated around the pivotal role of an alliance broker. This is seen as an important concept and role in the alliance business model. The question is:

> How can the single point of entry that large companies offer to clients and the overall management structure they represent be replicated in a peer regime?

This lies within the novel role of broker, an SME that in any one perceived business opportunity is the 'first among equals'. But not always the lead organisation, sometimes simply their main role is the provider of services along with all other alliance partners. It is the broker that delivers the one-stop shop presence to clients and who manages (within the structures and rules of the alliance governance) the identification, composition and orchestration of the appropriate partners and their contributions.

A key technical requirement will be to provide tools and services that enable the broker to be the channel of communication between the alliance and the outside world of customer, regulator, third party supplier, funds manager etc. The broker would ensure that an integrated approach to addressing construction industry key challenges, including sustainability, is adopted. Such an approach coordinates across technical and policy solutions, integrates engineering approaches with architectural design, considers design decisions within the realities of building operation, integrates green building with smart-growth concepts, and takes into account the numerous decision-makers within the industry. Obviously, the capability and maturity of an SME to fulfil this role is questionable, and remains a real challenge given the nature of the construction sector. However, a number of open questions remain, including the following:

- How to market and configure the alliance to the outside world? What is the role and scope of intervention of an alliance broker?
- What are his/her responsibilities and authority in relation to other alliance members throughout the entire product/service lifecycle?
- What business and organisational methods offer innovative and sustainable services throughout the collaboration?

12.7 How can an alliance be sustained?

The question of alliance sustainability raises issues related to risk sharing, motivation, discipline, trust and liability, and are summarised in Table 12.2.

The message is extended in Table 12.3 highlighting factors that are essential

Table 12.2 Criteria for sustainable alliances

Criteria	Current Situation	Expected Progress
Nature of business relationships	Short-term, local, ad hoc, fragmented and temporary business relations	Integrated and long-term partnering organised in virtual SME x within
Client focus	Project-oriented industry; each project is one-off production	*Process-oriented and performance driven* services in SME alliances
Investment strategy	SMEs focused on 'short-term cost thinking'	*Holistic lifecycle* (responsibility) and *cost-benefit approach*
Responsibility	SMEs tend to adopt a strategy of escaping responsibilities	Approaches centred on *long-term alliance-based guaranteed responsibility* and satisfaction
Competitiveness	Low	*High*: complementary expertise through alliances offering fully integrated solutions
Performance and workmanship	Low quality in timing (delays) and delivered solutions (failures)	*Guaranteed high performance* through continuous *learning and business improvement*
SME Profile and mode of operation	SMEs tend to employ and operate with traditional skills and little impact	New profession and skills for *sustainability, adaptability, resilience and health/safety*
Delivery	SMEs tend to focus on installations only	*Full lifecycle approach (responsibility)*: SME alliances to extend to *facility management and service guarantee*
Driving force for construction process	Demand-driven	*Supply-driven*: Suppliers are able to provide proactive and prompt solutions

to, or derive from, an alliance. For instance, currently in a construction project there is often little advanced warning of involvement; engagement may last only for a few months; there is often no integrated approach to ICT; contracts assume an adversarial attitude etc. Being in a long-term alliance(s) where partners work together permits better process integration using methods and tools that ease interoperability, improve performance, handle problems in a non-adversarial fashion etc, and hence sustain the alliance.

These provoke the questions:

• What holds an alliance together, what makes it last and what makes it successful?
• Do traditional change control mechanisms remain applicable in virtual SME alliance environments?

Trust is an essential factor, but there must be mechanisms that regulate working – set the ground rules.

- What might the benefits and incentives for SMEs be to work in this new way where contracts are primarily between themselves (criss-crossing) rather than spokes (radials) between SMEs and a controlling hub with its indirect and very often slow communication?
- How will the responsibilities and liabilities be managed in fluid situations between SMEs?

12.8 Role for information and communication technologies

It is widely acknowledged that the technological infrastructure necessary to support virtual business operations is now readily available (Rezgui, 2007a; Rezgui *et al.*, 2010).

However, SMEs, depending on their core business areas of activity, need dedicated software applications (Rezgui, 2007a). The involvement in long-term partnerships with other SMEs in the context of alliances requires a shift from a software-focused integration approach to a total lifecycle process integration philosophy (Rezgui, 2007a; Rezgui and Miles, 2009).

Moreover, existing ICT solutions tend to be technology (ICT) driven and are less sensitive to the socio-cultural and organisational issues that underpin long-term collaborative ventures. SME characteristics and needs have to be factored into any collaborative platform for the concept of alliances to work. This platform must be able to transcend traditional methods based mainly on stable/static business models to facilitate dynamic alliances. Information technology has been identified as a key enabler for partnering and alliance working (Latham, 1994; Egan, 1998). Table 12.3 contrasts the current situation with envisioned developments.

A number of open questions remain, including the following:

- How can integrated solutions be produced in appropriate situations that collaboratively combine and exploit the multi-disciplinary dimension of legacy and commercial tools used by SMEs?
- How to deal with 'work division' (decomposing into allocated tasks) 'work industrialisation' (automation in tasks), and 'work connection' (bridging between tasks elements of collaboration) – all with adaptation to SMEs?
- What is the necessary vision and systemic thinking required to achieve take-up as well as short to medium-term impact?

12.9 Promoting innovation in an alliance

There is a wide divergence in present SME capabilities and there will be equally diverse future capabilities (Bougrain and Haudeville, 2002). This variation reflects the different roles of construction SMEs, some being hands-on tradespersons whilst others are in the design or manufacturing sectors. The potential benefits of technical and process innovation can only be realised through SME members continuously learning and developing capabilities.

Table 12.3 ICT current situation and envisioned developments

SME Focused ICT-based engineering and manufacturing services

Criteria	Current situation	Envisioned situation
ICT adoption and diffusion	Low technology adoption	SMEs operating with *integrated/advanced ICTs promoting process innovation*
ICT provision model	Mainly based on licensing (expensive)	ICT service rental/pay-per-use (affordable)
ICT solutions for SMEs	Discrete, focused on business/engineering applications with little lifecycle/integrated approach	Integration of *software components* (services) to promote *SME long-term alliance building*
Project-based collaborative solutions	Limited, mainly provided through web-based document management systems or project intranets	SME *alliance portal* with a myriad of *integrated business and engineering applications/services*
Knowledge orientation	Focus more on data/products (IFCs) and less on services.	*Enhanced knowledge usage and best practice adoption* for innovative business processes
Socialisation and trust building	Ad-hoc communication through email and instant messaging	*Long-term community building* based on best practice and knowledge sharing
Tools to support communication and conflict resolution	Ad-hoc, based on traditional methods	Methods and tools for involvement and participation in *decision-making, and modelling/optimising* return on investment.

Individual skills are the route to achieving this. Individuals will need the skills to adopt and adapt to the new working practices. However, SMEs tend to have limited resources and limited time to dedicate to training. There is a need for a culture supportive of learning and change through continuous education. This would promote knowledge sharing between alliance members hence creating a more knowledgeable workforce. This produces a flexible organisation where people are more inclined to accept and adopt new ideas and changes through a shared alliance vision, as confirmed in (Powell *et al.*, 2004).

In the context of SMEs, given their inherent nature and structure, training and learning is best addressed through SME associations or federations where knowledge, best practice, training and learning materials are shared (Rezgui and Miles, 2010).

While an alliance has a strong operational dimension, a dedicated SME association should have a strong vocational dimension and needs to direct the learning and training process (making sure that documents, learning materials are up-to-date and directed to the right employees at the right time of the learning/training process).

However, further research is needed to understand:

- How SMEs perceive innovation?
- What factors are necessary to activate the knowledge sharing and creation capabilities of SMEs?
- How best to aggregate knowledge in an accessible way that is also maintainable?
- How SMEs mature and develop capabilities that sustain effective training underpinned by community-based learning and development?

12.6 Conclusion

The chapter argues that the way forward for improved knowledge management and capability building for the construction industry is for SME alliances to be established, forming knowledge-infused virtual networked organisations.

An alliance is a grouping of partners built around the principles described in the chapter underpinned by the following key features:

(a) A business model that provides sound controls whilst facilitating flexibility and innovation at the appropriate times. Legal, contractual and cultural diversity issues taken fully into account;
(b) A set of clearly defined but customisable roles (and in particular the role and functions of the broker);
(c) A total lifecycle process-driven philosophy where the intervention of each SME in the alliance complements and provides added value across the supply chain while delivering customised services and/or products to customers and clients;
(d) A performance-driven and environmental friendly approach to addressing clients' and customers' requirements;
(e) Technology and ICT mechanisms, underpinned by a sound legal and contractual framework, to support the operation and collaboration needs of the alliance (though appropriate to alliance, partner and project circumstances and needs).

This chapter has also identified a number of gaps and research questions that need addressing to ensure effective adoption of knowledge-infused alliance modes of operation. These research questions should form the focus of future research to provide further insights into SME alliance modes of collaboration.

The following chapter explores the ingredients for a successful construction organisation.

13 Ingredients for a successful knowledge construction organisation

The complex dimensions of a modern knowledge organisation; The acceptance of 'change'; Engaging stakeholders; Raising employees' awareness; The learning and training imperative; Building trust and confidence; Understanding corporate business processes; Understanding skills gaps; Planning and organising user-centred education and training; Continuous feedback and evaluation; Conclusion.

13.1 The complex dimensions of a modern knowledge organisation

Organisations are essentially 'social arrangements for achieving controlled performance in pursuit of collective goals' (Huczynski and Buchanan, 2001). Whatever the goals of the organisation (project management, production of building materials, creation of building designs, construction of buildings), the arrangements necessary for achieving controlled performance will involve the management of certain constant elements (Rezgui *et al.*, 2005), including: people (staff), technology (systems), tasks (skills) and structure (Leavitt, 1965). Other organisation theorists, such as Peters and Waterman (1982) have identified further elements of the organisation which they have incorporated into what is known as the 'Seven S' model – adding strategy, style and shared values to the four elements described by Leavitt (1965). These elements are all inter-related so that changes to any one element will have knock-on effects on the others. The elements must all operate in harmony together for the organisation to operate effectively.

For construction organisations to engage into new forms of collaboration will require consideration of the same combination of elements as for the management of 'traditional' organisations. However, managing and controlling the different elements so that they operate in harmony poses particular management problems in an alliance setting given the inter-organisational dimension of the alliance. Evidence suggests, for example, that in such circumstances management teams can find their control declining, which can generate further complications given the bureaucratic nature of these traditional organisations (Rezgui *et al.*, 2005).

The culture of an organisation (shared values, in the 'Seven S' model above) represents the 'cement' which holds an organisation together. A strong, i.e. clear and widely shared, organisational culture is now widely claimed to be critical to organisational success (Peters and Waterman, 1982).

The culture of an organisation can be defined as 'the collection of relatively uniform and enduring values, beliefs, customs, traditions and practices that are shared by its employees, learned by new recruits, and transmitted from one generation of employees to the next.' (Huczynski and Buchanan, 2001). Organisational cultures derive from basic assumptions and values held by key individuals in the organisation (e.g. the founder, the CEO, the senior management team) which shape organisational objectives, structures and processes, and which are promoted and shared both directly and indirectly through a range of mechanisms (Rezgui *et al.*, 2005). Developing and promoting a corporate culture is a challenge for a construction alliance, as the culture should be embraced by the constituent members of the alliance.

Whilst construction alliances are enabled via existing and emerging technologies, they remain principally human constructs. The success of the alliance relies on its capability to operate synergistically while creating and sustaining value for its employees across the alliance (i.e. participating organisations). In this context, the management of the human capital of the alliance, and more generally, its 'intangible assets' play a determinant role. Thus the socio-organisational 'equation' of a successful alliance consists of a combination of technology, culture and organisation, in which issues including trust, confidentiality, knowledge policies, etc. must be blended successfully toward the shared alliance purpose (Rezgui *et al.*, 2005).

13.2 The acceptance of 'change'

The new economical and financial climate creates increased pressure on organisations to embrace change to remain competitive in a fierce and uncertain business environment. The survival of organisations will depend upon their ability to respond and adapt positively to the challenges posed by this new environment. Whilst technology plays an important role in enabling the right modes of collaboration and business models, including alliances, to respond to these challenges, it has been shown that the underpinning solution has a strong socio-organisational dimension. There is a need for employees within organisations to accept change, and for managers to promote a culture supportive for change (Rezgui *et al.*, 2005).

There are four main types of change strategy, which can be summarised on a 2×2 matrix as shown in Figure 13.1 (Rezgui *et al.*, 2005). The strategies vary from the most radical – where stakeholders stop using a process or product (e.g. management information system) one-day and switch to a fully new process or product the next, to the most evolutionary – where stakeholders progressively make the change over a period of months. The greater the scale of change at one time, the more difficult it is for the stakeholder to adapt. However, there is no 'right' way – management must choose the one most appropriate for the organisation, bearing in mind the organisational culture and context. Each of

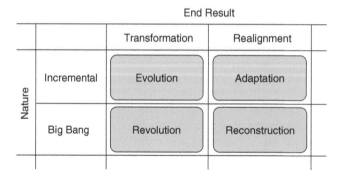

Figure 13.1 Types of change (adapted Rezgui *et al.*, 2005).

the different strategies has different advantages and dissadvantages, and different implications for the likely time and resource requirements for the change programme. However, it is possible to use a combination of strategies during the course of a programme.

While the transition from 'traditional' organisations towards 'virtual' organisations or alliances can be experimented by real construction projects that involve parts of the alliance members, when organisational boundaries are being crossed the nature of the changes required internally may be quite large. Issues of trust as noted earlier are paramount, ensuring that the right knowledge policies and systems are put into place. These should enable the virtual alliance to conduct its business in a serene environment where information and knowledge is shared and nurtured, but at the same time protecting and preserving the intellectual property rights of constituent organisations and members, and confidentiality of other organisational knowledge.

It will be necessary to re-think and create management and leadership structures that apply to the alliance, although these may not be in harmony with existing structures within the 'real' organisations. Similarly, communications processes will need to be established and working styles, roles and rewards may need to be moderated for the 'virtual' alliance to work effectively. Changes that benefit the members of the 'virtual' alliance may cause resistance or discontent amongst members of the organisations who are not directly involved in the alliance business and this will need careful management.

Senior management must therefore adopt a strategic approach to change management, taking into consideration issues not only at the alliance level but also at the parts of the 'real' organisation that are likely to be affected and ensuring that the impact of the change is managed in both.

The emerging issues related to change that need to be considered include the following (Vakola and Wilson, 2004):

- Partnering;
- New commitment to lifelong learning;
- New knowledge infrastructure providing just-in-time information delivery;
- Much better use of corporate knowledge resources;
- A new willingness to learn from experience, recognise mistakes and reward good practice;
- Industry-wide knowledge sharing through benchmarking and electronic knowledge pools;
- Total lifecycle orientation. Including information transfer through the supply chain.

In attempting to cope with change processes, it is important for construction companies to examine available tools, methodologies and training materials in order to design, implement and evaluate a change programme. Tools, models and methods of change are similar in concept and present many commonalities (Senge, 1990). Beckhard and Harris (1987) analysed a three-phase model indicating that complex organisational change can be conceptualised as a movement from the present state to a future state. The most important phase is the transition state which analyses how the company will move from the current state to the future one. As a result, change is a matter of assessing the current situation, developing the appropriate strategy and vision and designing the change process.

13.3 Engaging stakeholders

Academic literature suggests that for a planned change to be successful, it is crucial that those who will be affected by it (i.e. employees and the stakeholders) are engaged in the planning of that change (Bennis *et al.*, 1985). This does not mean that all stakeholders have to be involved in every aspect of the change programme, but it is essential to identify key stakeholder inputs (requirements, concerns etc.) and address these if the change is to be effective. There are also significant benefits to be gained from enabling stakeholders to actively participate in a change project, compared to simply communicating to, or seeking information from them. By investing time and effort in the change process, stakeholders begin to identify with and to develop personal goals for the planned change. The engagement process also gives early warning of potential difficulties, which often enables solutions to be found before significant adverse impact occurs (Rezgui *et al.*, 2005).

In order to engage stakeholders and to understand the nature of their 'stake' in a change project, it will be necessary to establish who the stakeholders are, what roles they hold in relation to the change (in particular their relative power and influence), how they are likely to accept the change (this will be influenced e.g. by their current concerns and their expectations about how the change will affect them), and what requirements they have in relation to the change.

The stakeholders will involve the members of the planned construction alliance, their existing and future clients and customer base, local/regional/national authorities with a stake in the planned alliance, etc. Investment in time during this consultation stage will help establish the right foundation on which to plan and build the future of the planned alliance.

13.4. Raising employees' awareness

Several factors have led to important deficiencies in education and training in the construction sector. Various surveys and studies revealed a lack of awareness in the role of knowledge and ICT within senior management and decision-making circles and of their potential benefits in improving productivity and competitiveness (Vakola and Wilson, 2004). In fact, business processes tend to be driven by cost as opposed to performance and value, with little importance given to intangible aspects such as human capital. Furthermore, decision-making with regard to ICT and investment in knowledge initiatives across projects and organisations is often influenced almost exclusively by the desire to lower costs. A few important and influential clients have already understood the potential of knowledge management and related learning and training and have started reaping off the benefits of their investment (Rezgui, 2007b). However, these tend to be a minority and cannot in any way be generalised to the industry.

It has been shown that the nature of the culture in place within organisations is critical to the success of any change or process improvement initiative. Characteristics from the 'participatory' type of culture (including flat structure, open communication channels, participation and involvement in decision-making) enhance sharing of information and facilitate process improvement initiatives. 'Bureaucratic' cultures cannot support effectively these cross-functional teams and as a result, they can hinder the change process.

It is important to promote education and training awareness to company staff, including senior managers and decision-makers, and address the barriers that are preventing them from adopting the right attitude towards knowledge sharing. This can be facilitated and implemented in a variety of ways, using a variety of instruments that have to be devised on a company per company basis. This includes face-to-face meetings (in particular with senior staff and decision-makers), group workshops involving team managers as well as general staff, wide distribution of newsletters and brochures (adapted to take staff profile into account) showcasing and illustrating the business process benefits of knowledge technologies and related education and training to improve productivity and competitiveness.

The challenge is to get key staff (including senior management) adhering to the vision of the learning organisation. A learning organisation is an organisation that has a culture supportive for learning and has a mission to drive change through continuous education and learning of its staff. It promotes knowledge sharing between employees hence creating a more knowledgeable workforce.

This produces a very flexible organisation where people are more inclined to accept and adopt new ideas and changes through a company globally shared vision. Learning organisations as argued by Senge (1990) are being met by the following ingredients:

- Personal mastery – the capacity to clarify what is most important to the individual;
- Team learning – based on a dialogue in which assumptions are suspended so that a genuine thinking together occurs;
- Mental models – the capacity to reflect on internal pictures of the world to see how they shape actions;
- Shared vision – the ability to build a sense of commitment in a group based on what people would really like to create;
- Systems thinking – the capacity for putting things together and seeking holistic solutions.

Individuals' learning preferences can easily be recognised in the above five ingredients (Senge, 1990). Every person learns in a specific way, either individually, in teams, via shaped models, etc. An understanding of the above by the entire organisation staff is a prerequisite to successful implementation of the learning organisation vision that engages all its employees.

13.5 The learning and training imperative

Training remains an important issue that is under-addressed in the construction sector, and when available it is organised on a vocational and/or voluntary basis (i.e. undertaken for reasons of continuing personal development). Individuals in the workforce need to continue to learn, update and add to their skills in order to remain employable, have the right skills to address current industry challenges, and thus add to the competitiveness of the construction sector.

The construction sector faces many problems with people development because of continuous change in working practices, new technological developments, lack of training and learning, and poor people management (Vakola and Wilson, 2004). Lifelong learning is considered as a critical success factor in order to support cultural changes, and has the potential to attract and retain young and qualified people to the industry.

There is growing recognition in the construction sector that the potential benefits of technical and process innovation can only be realised through individuals at all levels learning and developing a considerable array of new capabilities. There are many drivers for change in the sector (Lowe, 2001). These include the trend towards new contractual arrangements such as partnering, process improvements initiatives to improve information flows through the supply chain, information management and knowledge sharing for distributed teams, and enhanced communications within and between teams – often across organisational boundaries.

The required capabilities will not, in the main, be developed through traditional training courses but through the creation of well-planned learning opportunities in a supportive learning environment.

The imperative to transform organisational culture and working practices has profound implications for individual and organisational learning. For example, while partnering arrangements offer significant benefits to all stakeholders, these will only be realised fully if individuals such as the supervisors, construction engineers and site workers understand the concept and the implications of the new arrangements: thus individuals accustomed for most of their employment history to simply following instructions must migrate from a largely passive role to an active role where they take responsibility (planning, anticipating requirements, troubleshooting, communicating, monitoring and reporting etc). This represents an enormous challenge and will require an integrated change strategy for learning and training at corporate and at sectoral level.

Establishing vertical and horizontal skills alliances within and across different disciplines in construction industry education and training could foster increased collaboration and innovation for the future (Vakola and Wilson, 2004).

As discussed earlier, construction partnerships, like strategic alliances in general, are less natural learning entities than are single organisations (Larsson *et al.*, 1998). The reasons should be understood and addressed before planning training and learning are addressed. Hence, a particular attention should be given to the dynamics of these alliances.

13.6 Building trust and confidence

Teams with trust converge more easily, share knowledge effectively, organise their work more quickly and manage themselves better (Lipnack and Stamps, 2000). Trust is essential in construction virtual teams due to the complexity of the project environment, as discussed earlier. It is vital that project teams have the ability and understanding of how to work together to solve problems and this is easier when a high degree of trust exists. The importance of understanding the need to build trust and confidence is a matter for all stakeholders from senior managers taking a strategic perspective to those responsible for daily businesses and project operations, and one way to ensure it is understood is to include such issues in education and training initiatives.

Research indicates that people generally tend to trust people rather than companies and that trust ultimately emerges where communicated information is reliable, people stand by their promises and outcomes equal or exceed expectations (Rezgui *et al.*, 2005).

In a virtual setting, trust and confidence must be promoted through suitable knowledge and collaboration technologies which have been implemented, tested and validated in a real-world context. Another aspect of trust relates to technology introduction. The fact that new ICT tools are continuously introduced in the workplace to accomplish tasks does not change the fact that we must rely on people to engender trust. Indeed people work together because they trust one another and successful virtual teams pay special attention to building trust (Lipnack and Stamps, 2000).

13.7. Understanding corporate business processes

Any business process improvement initiative should build on a comprehensive understanding of inter- and intra-organisation aspects, including core business processes that help understand the complex environment in which the organisation operates. The key objectives are to:

- *Understand current company strategy:* In what direction is the company going (what are their likely current/future strategic ICT requirements?).
- *Understand the structure of the organisation:* Division of the work, the tasks and the responsibilities both horizontally and vertically.
- *Understand the culture of the organisation:* Values, norms and views shared by employees. These may be expressed in the form of symbols, stories and legends.
- *Understand the systems in use:* Rules, procedures and guidelines, software/hardware systems in use.

This analysis should be facilitated through the use of internal consultants. Internal consultants will be less likely to encounter resistance and will be more respected by employees. This will help secure ownership of the learning and training process, which should help to ensure management support.

The principles of action research (defined as being responsive, cyclic, participative, qualitative and reflective) can be used during this business process mapping exercise and should help reveal past problems in process improvement initiatives (including ICT deployment initiatives) within the organisation. Similarly, reactions from employees can uncover cultural aspects of the organisation which may assist or hinder any knowledge and learning initiative/strategy. It is important to understand and control these parameters during the business mapping exercise.

13.8 Understanding skills gaps

Organisations need to analyse and map available skills within across their departments and subsidiaries. An approach making use of a combination of a process- and user-driven analysis can be adopted. The process-driven analysis will build and extend the work involved in understanding and mapping corporate business processes and attempt for each process component (including activity/task) identify company staff that enable, control, and perform the activity/task. The user-driven analysis will establish a profile of each employee and highlight current qualifications, roles, responsibilities, core competencies, areas of expertise, past and ongoing achievements and experiences (e.g. projects), etc.

Roles have responsibilities attached to them and can be discharged by one or several persons. It is important to understand the skills that are required to perform a role and ensure that individuals who are in charge receive the right training to perform their duties. In that respect, roles need reviewing and creating to ensure that learning and training strategies get adopted and implemented.

This will provide a general picture about the capabilities and ICT literacy of the company staff, which can give interesting indications about the level of maturity of the business processes described in the previous stage.

There is growing tendency for business processes to be partially or fully supported by information systems. This is likely to affect the majority of the task force in the sector, which will require a general level of 'computer literacy' skills. These computer-supported business processes concern the majority of stakeholders in the sector and span the whole project lifecycle. There are obvious skill gaps in the sector due to a weak ICT literacy from employees and lack of education and training support within most organisations. As mentioned earlier, training courses to deliver the above skills tend to be vocational and/or voluntary (i.e. undertaken for reasons of continuing personal development). It is argued that a more strategic approach should be taken.

13.9 Planning and organising user-centred education and training

Learning is an individualistic characteristic based on individual's culture, habits and behaviour. Training on the other hand is regarded to be in the domain of companies and institutions with a purpose as argued by Sloman (2001) of 'invention designed to improve the knowledge and skills of employees'.

People think and act in a way that is conditioned by what they have learned. For organisations to offer employees learning opportunities can be of a strategic value. Honey (1998) has argued that the following affirmations can be made about learning:

- Learning is both process and an outcome.
- Learning is not just about knowledge. It is also about skills, insights, beliefs, values, attitudes, habits, feelings, wisdom, etc.
- Learning outcomes can be desirable or undesirable for the learner and for others.
- Learning process can be conscious or unconscious.
- Conscious learning process can be proactive or reactive.
- Learning process occurs inside the individual, but making the outcome explicit, and sharing them with others, adds value to the learning.
- There is no right way to learn for everybody and for every situation.

Individuals need the skills to embrace a knowledge culture, use technology and adopt and adapt to new working practices. There is a need to encourage employees 'to work smarter not harder'. This can be achieved in most organisations through identifying people who are interested in finding better ways of doing things and giving them opportunities to explore options and to introduce new methods and systems on a carefully monitored trial basis, as a preliminary to wider roll-out.

13.10 Continuous feedback and evaluation

Organisations need to put in place mechanisms to review the efficiency and effectiveness of knowledge initiatives, and measure the business benefits achieved. In order for an organisation to perform an evaluation process, success factors must be clearly defined. This involves monitoring and reporting results to top-level management, presenting learning and performance gains in enhancing productivity and competitive position on the market.

However, organisations need to be aware of the fact that evaluations are costly and that a key part of planning is to choose appropriate sources of information and methods of collecting evidence based on well planned metrics. But how can an organisation select the best method to evaluate a knowledge transformation? There is a variety of methods available and this fact alone indicates that no single methodology is 'the best'.

Broadbent (2002) suggest diverse approaches concentrating on participants' experience (including managers, developers, instructors and learners). He suggests an evaluation according to the change in their attitudes and skills preceding the course, during the course and at the end of the course by applying methods based on questionnaires and focus groups.

This is critical, as failing to demonstrate returns on knowledge, learning and training investments can hinder the knowledge and learning programme initiatives.

13.11 Conclusion

This chapter discusses important issues related to knowledge and learning in construction organisations. These issues tend to be overlooked in construction IT research, dominated traditionally by technology-led developments. The chapter explores the social, organisational and technology aspects of a construction alliance and highlights the issues that need addressing in order to negotiate the necessary transition from a traditional to a knowledge-driven organisation that engages effectively in knowledge-driven alliances.

14 Value creation

The future of knowledge management in construction

Reflecting on knowledge management in construction; Why value creation? Role of human networks in value creation; Role of social capital in value creation; Role of technology in value creation; Role of intellectual capital in value creation; Role of change management in value creation; Concluding remarks.

14.1 Reflecting on knowledge management in construction

We have in Chapter 4 discussed and categorised knowledge management in construction in terms of three generations.

The first generation adopts a document-based approach to KM. This is reflected in the initial adoption of electronic document management systems (EDMS) in some leading organisations and on large-scale projects in the early 1990s. We have discussed the fact that some form of knowledge is embedded in documents (such as standards, regulations, full specifications, etc.) but that this requires human decoding and interpretation (section 6.10). Often, the context of knowledge is not captured but lost, which hinders proper knowledge elicitation. Documents are managed and archived using techniques drawn from library sciences, mainly based on pre-defined keywords. This generation of KM is driven by business process automation through technology (IT). These document-centred knowledge systems tend to overlook the socio-cultural aspects that underpin knowledge management activities. It has been argued earlier in the book that KM has a strong social dimension and that KMS should promote communities of users sharing explicit and tacit knowledge. Three factors play a key role in the successful re-use of knowledge shared and created in a knowledge space (Markus, 2001):

(a) the costs involved in creating and using entries;
(b) the incentives people have to create and use entries, and;
(c) the roles of intermediaries in the creation and maintenance of shared knowledge spaces and the facilitation of their use.

These factors explain the relatively low impact and uptake of document management systems in construction. These tend to be successfully adopted in organisations with a supportive culture and on projects with large client organisations that can afford to invest in these technologies.

The second generation of KM in construction is characterised by efforts aimed at (a) knowledge codification (including best practice management) and (b) conceptualisation of buildings through building information models (BIM). These efforts have been widely encouraged by advances in computer aided design (reflected in recent parametric and BIM-oriented CAD software) and the advent of the semantic web. This second generation of KM involves the proliferation of advanced search facilities using semantic text and knowledge mining techniques, based on dedicated taxonomy/thesaurus/ontology, as well as the emergence of discipline-oriented knowledge systems reflected in some early forms of decision support systems. This generation is characterised by an increasing social and process awareness about IT adoption and deployment. In a nutshell, there is a shift from intra- to inter-company capability development, with a focus on projects.

Value creation from data, information and knowledge is what characterises the future of knowledge. In fact, the industry is now influenced and driven by highly demanding clients and continuously changing regulatory requirements to meet higher environmental and comfort standards. Moreover, clients demand greater flexibility to satisfy larger customer choices, and smart buildings that ensure maximum comfort and safety to the occupants. The best illustration of this transition is reflected in in-house sustainability and health and safety initiatives in some leading construction organisations.

Sustainability goals can only be achieved if existing and new resources of knowledge and expertise inform construction activities. Some of this comes in the form of good practice and standards, but much comes from situated and contextual appreciations of sustainability goals and local practices developed within and across projects and organisational boundaries. In this respect, knowledge-sharing initiatives provide means to capture, represent and disseminate sustainability information and experiences acquired on projects, and enable these to be nurtured within and across organisations and applied successfully in projects with real potential for impact.

Similarly, in response to the critical situation of health and safety on construction projects and the consequences in terms of image and reputation to the industry, construction companies have started using basic knowledge management solutions to record incidents on construction site and make these widely available and accessible to employees through company intranets (Rezgui, 2007b). There is an increasing awareness that the capture, representation and dissemination of health and safety measures create real values in terms of reduction of accidents on sites and improvement of the overall well being of staff and future occupants.

These forms of value creation and management are what characterises the future of knowledge as elaborated further below.

14.2 Why value creation?

Impact and value creation are what should drive modern organisations. There is a strong move in industry to deliver higher-value buildings that address current and emerging environmental and societal concerns. Clients demand better adaptability and resilience of their buildings.

Leading organisations are now considering their corporate social responsibility which is often now factored into their corporate strategy. There is also a greater awareness about the role of 'intangibles' in sustaining innovation and competitiveness. There is a gradual shift from 'tangible' to 'intangible' assets in the way the organisation is managed and valued. We are now entering a generation of knowledge maturity with a higher sense of responsibility from all stakeholders and a firm intention to learn from the consequences of decades of ill-practice in the industry.

The relationship between value creation and KM has been discussed extensively in recent literature (Chase, 1997; Despres and Chauvel, 1999; Gebert et al., 2003; Liebowitz and Suen, 2000; Rezgui, 2007a), and it has been argued that KM processes have inherent value creation capabilities that ought to be better exploited (Gebert et al., 2003). Value creation takes place and is facilitated by (a) creating knowledge repositories, (b) improving knowledge access, (c) enhancing cultural support for knowledge use, and (d) managing knowledge as an asset (Davenport and Prusak, 1998).

The quest for value creation and innovation in industry has given a new impetus to KM research (Aranda and Molina-Fernandez, 2002; Huseby and Chou, 2003). As noted earlier in the book, KM is perceived as a framework for designing an organisation's goals, structures, and processes so that the organisation can use what it knows to learn and create value for its customers, community and society (Choo, 2000; Rezgui et al., 2010).

As alluded to throughout the chapters of this book: 'Value creation is grounded in the appropriate combination of human networks, social capital, intellectual capital, and technology assets, facilitated by a culture of change' (Vorakulpipat and Rezgui, 2008).

In fact, the concept of community of practice (Wenger et al., 2002) was introduced as an effective social activity to share and nurture tacit knowledge within and across projects and organisations. This helped managers to understand, and it raised their awareness about, the role of *human networks* in motivating employees to share and create knowledge.

In this context, the *social capital* of an organisation emerges as an essential ingredient to help employees develop trust, respect and understanding of others. Because of its (social capital) emphasis on collectivism and co-operation rather than individualism, distributed community members will be more inclined to connect and use electronic networks when they are motivated to share knowledge (Huysman and Wulf, 2006; Rezgui 2007c). KM environments may foster *social capital* by offering virtual knowledge spaces for interaction, providing the context and history of interaction, and offering a motivational element (e.g. reward) to encourage people to share knowledge (Huysman and Wulf, 2006).

The *intellectual capital* of an organisation encompasses organisational learning, innovation, skills, competencies, expertise and capabilities (Rastogi, 2000). Liebowitz and Suen (2000) suggest a strong and positive relationship between the value creation capability and the intellectual capital of an organisation, pointing to factors such as training, research and development investment, employee satisfaction and the development of relationships

Focusing on *technology*, the majority of KM initiatives involve, to a lesser or greater degree, information technology (IT). Technology, including KMS, is an essential means of promoting and creating value out of knowledge in the distributed and fragmented networks of the construction industry.

Lastly, *change management* plays an increasingly important role in sustaining 'leading edge' competitiveness for organisations in times of rapid change and increased competition (McAdam and Galloway, 2005; Reddy and Reddy, 2002; Wheatcroft, 2000). The survival of organisations will depend upon their ability not only to adapt to, but also to master, these change-related challenges.

The sections below discuss the role of each of the above factors (human networks, social capital, intellectual capital, technology assets and change management) in promoting and sustaining value creation.

14.3 Role of human networks in value creation

A collectivist or participatory culture is needed in a team to create sustainable network ties between team members. Knowledge and its value dimension can only be exploited if they are cultivated across the network of relationships of individuals within teams, projects or their organisation. Employees need the right environment (including a collective and participatory culture) in order to engage in problem-solving and decision-making activities and to offer a range of different skills, abilities, knowledge, and experience to ensure that creative ideas are supported and value is created. However, a number of issues can hinder the development of effective human networks in organisations and across projects.

There is a need to develop within teams (across organisations and projects) strong communication and collaboration protocols, including codes of conduct, standards for availability and acknowledgement. The importance of a shared project knowledge base in a virtual team context has been discussed earlier. There is strong agreement that shared project knowledge improves communication and cohesion amongst team members and promotes shared language and mental models across teams, leading to the development of trust and a culture of knowledge sharing (Crampton, 2001; Rezgui 2007a; Suchan and Hayzak, 2001).

However, potential clashes of cultures on projects (due to their multinational and multi-disciplinary dimensions) are an issue that has not been addressed in the construction industry. There is a need for the goals of the organisation(s) and project to be shared and embraced collectively. It has been noted that differences in organisational affiliations on construction projects can reduce shared understanding of context and can inhibit a team's ability to develop a shared sense of identity (Espinosa *et al.*, 2003).

Virtual modes of collaboration may reduce face-to-face interactions during the virtual team lifecycle and in particular during the inception stage where the vision, mission and goals should be communicated and shared (Rezgui, 2007a). Team collaboration through face-to-face communication creates stronger social relationships. However, these are difficult to establish in virtual contexts due to a number of factors, including the lack of emotional expression. There is a greater acceptance of face-to-face rather than virtual interaction amongst construction practitioners. Virtual communication such as email may form bridges between team members (e.g. across different geographical locations) but this may not yet provide the right bond to sustain robust socially constructed relationships. Construction relies heavily on ad-hoc relationships between individuals and companies based on a rapport of trust. This is an important project success factor. Trust among employees can promote a knowledge-sharing culture and is important for the exchange of knowledge: 'without trust there is no knowledge sharing' (Lee, 2001; Roberts, 2000; Sveiby, 1999).

The team-building exercise is overall essential in order to establish a clear team structure and shared norms (Crampton, 2001; Tan *et al.*, 2000). Early face-to-face interactions between team members during the team's launch phase tend to (a) improve the team's project definition (Sarker *et al.*, 2001), (b) promote socialisation, (c) engender trust and respect among team members (Crampton, 2001), and (d) enhance the effectiveness of subsequent electronic communications (Maznevski and Chudoba, 2001).

It is only when these factors are met that true sharing of best practice, collaboration in problem solving, and nurturing of knowledge can take place, and ultimately, sustained human networks established.

14.4 Role of social capital in value creation

In terms of innovation and competitiveness, trading and sharing of knowledge have become increasingly important and have forced organisations to create market spaces and places to promote knowledge sharing activities (Choo, 2003). Interaction or conversation between employees, for example, is often perceived as the simplest approach to transferring knowledge within an organisation. Nevertheless, it may be inconvenient where cultural barriers exist (Davenport and Prusak, 1998). It is argued that to align knowledge sharing with organisation culture, designing and implementing KM to fit the culture can be more effective than altering and changing the culture itself (McDermott and O'Dell, 2001).

Organisational culture involves two dimensions: the visible dimension – 'thing', and the invisible dimension – 'seen but unspoken' (McDermott and O'Dell, 2001). McDermott and O'Dell maintain that organisations ought to make sharing knowledge visible to solve practical business problems by, for example, making it directly part of the business strategy, initiating it obliquely on to another key business, routinising, matching the organisation's style and aligning reward. They repeatedly emphasise that databases, knowledge systems and knowledge initiatives need to have a clear business purpose so that organisations may easily understand how sharing knowledge contributes to business goals. In fact, the reason KM programmes fail may be attributed to a lack of a clear connection with a business goal.

The central premise of social capital is that social interactions between individuals and groups through mechanisms such as networks, shared trust, norms and values can be used to achieve various mutual benefits. Social capital is identified as a key factor encouraging community cohesion, and fostering social and civic engagement.

Different types of social capital produce different outcomes. The concept of social capital has recently been adopted within the knowledge management community (Vorakulpipat and Rezgui, 2008). A focus on social capital in relation to knowledge sharing shifts the attention from individuals sharing knowledge to communities as knowledge-sharing entities.

Evidence suggests that the higher the level of social capital, the more communities are stimulated to connect, concert, share knowledge and experiences and agree on common courses of actions (Dale and Newman, 2008). A participatory culture helps develop trust, respect and understanding for others at different levels in the construction sector. Clearly, a culture of confidence and trust in organisations and projects are the foundation for effective social capital.

A culture that recognises tacit knowledge and social networks results in the promotion of open and constructive communication between staff, allowing them to develop sustained social links and share common context understandings and interpretations. The will to share tacit knowledge through social communications can be interpreted as a means to (a) break down barriers between employees and management, (b) establish stronger relationships among them, (c) allow employees to gain confidence and participate in decision-making tasks, and (d) practice and improve their knowledge-sharing and creation capabilities.

Ultimately, people choose to share experiences and best practice because they trust one another and successful virtual teams pay special attention to building trust throughout their collaboration. It has been discussed earlier that practitioners generally tend to trust their peers rather than companies and trust ultimately emerges where communicated information is reliable, people stand by their promises and outcomes equal or exceed expectations. Teams with trust converge more easily, organise their work more quickly and manage themselves better (Sarker *et al.*, 2001). These ingredients are essential to promote social capital which in turn provides a basis for value-creating knowledge activities.

14.5 Role of technology in value creation

Despite the tendency to emphasise the role of IT in KM, there is an increase of powerful arguments for a more holistic view which recognises the interplay between social and technical factors (Pan and Scarbrough, 1998). Therefore, a socio-technical approach to knowledge sharing emerges and is now applied in many organisations. This approach emphasises the interplay between the KMS and the organisational context. Management and leadership play a critical role in establishing the multi-level context for the effective assimilation of KM practices (Pan and Scarbrough, 1998).

There is a need for social-oriented communication and social forums to engage construction practitioners in knowledge-sharing activities in industry. There is a belief amongst managers that technology is a panacea that has the potential, on its own, to solve problems faced by practitioners. This perception leads to KM fallacies or traps that directly influence the perceived functionality of IT applications for the support of KM initiatives (Huysman and de Wit, 2002). These fallacies relate to the tendency of organisations to concentrate too much on the IT role supporting KM practices, especially knowledge sharing, resulting in the 'IT trap' (Huysman and Wulf, 2006). It is important to recognise that IT is not independent from the social environment, as it is not the technology itself but the way people use it that determines the role of IT in supporting knowledge management practices.

It has been argued earlier in the book that current ICT systems exhibit limitations in supporting collaborative working, as these do not integrate seamlessly with the engineering applications used on a daily basis on projects, and provide therefore limited support to the practice (Rezgui, 2007a). There is a large feeling in industry that ICT have an invasive nature and impose demand on individuals and teams to continuously adapt to their introduction on projects. Current solutions require constant adaptation and re-configuration of existing legacy/commercial systems, while offering limited growth path and scalability.

In fact, team members on projects are affected more by the newness of the technology being used than by the newness of the team structure itself (Rezgui, 2007a). These problems of technology adoption can have a negative effect on individual satisfaction with the team experience and performance (Kayworth and Leidner, 2000; Van Ryssen and Hayes-Godar, 2000). Conversely, when team members are able to deal with technology-related challenges, high trust develops (Jarvenpaa and Leidner, 1999).

A number of critical success factors for the adoption of KMS have been identified (Rezgui, 2006). These include:

- *Ease of use*: the solution must present a usable intuitive interface with, ideally, a standardised feel across all the available functionality or software solutions offered across the organisation or project. This is now possible thanks to advances in web services technology.
- *Adaptive*: the solution should be flexible and adaptable as the needs of the organisation and users change and mature. KM systems should have the ability to learn from their own use and user behaviour and to adapt to new situations while minimising manual maintenance, configuration and support. An appropriate representation of the history of knowledge-sharing

activities may be useful since it allows human actors to better understand and refer to past interactions (Huysman and Wulf, 2006).

- *Support existing practices*: the solution should, as far as possible, be flexible enough to accommodate existing practices while building on existing corporate solutions. If change is required this should be introduced in an incremental and evolutionary way while providing substantial benefits to the end-users as well as to the company management staff.
- *Open and scalable*: the solution should make use of open standards and technologies, avoiding 'closed' vendor-specific solutions.
- *Model and content-based*: the use of a construction-based ontology holding a shared understanding of construction concepts promotes the wide adoption of the solution while supporting the transition from a document- to a content-centred approach to KM.
- *Ambient access*: the solution should be accessible anywhere, anytime from a wide range of devices. A similar level of functionality should be provided regardless of time and distance.
- *IPR and security sensitive*: the solution must take into account contractual, legal, IPR (intellectual property rights), security and confidentiality constraints.

Controlled access to information and knowledge enhances trust in technology, an important factor in promoting a culture of knowledge sharing facilitated by ICT.

The lack of a clear vision and ICT strategy within projects and across companies in industry is a concern, as discussed earlier in the book. The prevailing policy based on acquiring off-the-shelf software solutions fails to deliver. These commercial solutions tend neither to accommodate existing practices nor build on existing corporate solutions.

Technology with its ubiquitous dimension is an essential ingredient for value creation. However, the organisation's success with the adoption of IT will not only depend on IT functionality and employees' skills, but the appropriate social context and culture that can absorb and benefit from these technologies (Zack and McKenny, 2000). This requires the use of socially embedded technologies or collaborative systems informed by the belief structures of employees, facilitated by management teams.

14.6 Role of intellectual capital in value creation

The argument has been made earlier in the book about the importance of 'intangibles', and more precisely, intellectual capital, in the innovation dimension of an organisation. Intellectual capital involves the factors/ingredients discussed in the previous chapter, including organisational learning, skills, competencies, expertise and capability development.

Organisational learning is an issue in the industry as shown in the previous chapter. There is a need for more adapted training and a more strategic approach to learning in organisations. Adapted training can foster cohesiveness, team work, commitment to team goals, individual satisfaction and higher perceived decision quality (Tan *et al.*, 2000; Van Ryssen and Hayes Godar, 2000; Warkentin and Beranek, 1999). However, short timescales in construction, due to simultaneous involvement in projects, create additional pressure and leave little time for strategic training and learning.

There is concern about the bureaucratic and hierarchical culture in large construction organisations, which is in several instances reproduced in teams. A wrong culture will act as a barrier to intellectual capital development. Intellectual capital should be nurtured within the social fabric of the organisation or projects. Issues related to trust, social cohesion, motivation and satisfaction are continuously raised as noted earlier, and the sense of a political climate that engenders mistrust and adversarial relationships is commonplace in the industry.

Ultimately, intellectual capital is what drives innovation, fuelled by employees and the knowledge they possess.

14.7 Role of change management in value creation

Technology introduction is one of the main drivers for organisational change in construction (Rezgui *et al.*, 2005; Vakola and Wilson, 2004). From the 'Seven S' model shown in the previous chapter, it is clear that technology, including knowledge management systems, are only one element of the organisational landscape, and that the introduction of new technology will undoubtedly impact on all other aspects of the organisation. Failure to proactively manage this impact is likely to result in a wide range of organisational problems. Conversely, the introduction of new technology offers opportunities to review other aspects of the organisation and to introduce new business processes, roles, responsibilities, etc., which will collectively enhance organisational performance. Sociotechnical systems theory (Choo, 2000) highlights the importance of designing social (organisational) systems and technical systems in parallel, in order to achieve optimisation of both.

Hence, change management includes technical and human issues. It has been shown that the invasive nature of ICT and the need to adapt to continuous technology introduction across projects is a concern for practitioners. These problems of technology adoption can have a negative effect on knowledge management initiatives and individuals/teams experience and performance. Conversely, when team members are able to deal with technology-related challenges, a culture of knowledge sharing can be promoted. In terms of human issues, knowledge value creation implies new approaches to the management of human resources, information and knowledge within organisations. While the potential gains can be easily articulated, the necessary changes might be resisted. Therefore, to be effective, any KM programme should be incorporated within a change management programme that takes into account the team-based structure and discipline-oriented nature of the construction industry.

The concerns about the necessary changes in the organisation should be raised explicitly, and addressed collectively.

14.9 Concluding remarks

Knowledge management is a complex and evolving subject. The book has argued the case for knowledge management techniques and approaches adapted to the characteristics of the construction industry.

The book has first explored the changing nature of work and organisations in construction and the gradual shift from tangible to intangible assets centred on knowledge and the intellectual capital of an organisation. Organisations worldwide are refocusing strategies to deliver new and existing products and services with maximum value to the customer at the lowest price. In this context, success lies in the ability of an enterprise to combine complementary and distributed expertise and skills to continuously innovate and remain competitive.

The role of knowledge in construction is then explored. The barriers to innovation are discussed, as well as knowledge needs of the various stakeholders involved in the design, construction and maintenance stages of a building.

The book then provides a critical and evolutionary analysis of knowledge management in the construction industry. An evolutionary KM framework is provided that presents three proposed generations of KM in construction, in terms of three dimensions that factor in (a) the capability of individuals, teams and organisations in the sector, (b) ICT evolution and adoption patterns, and (c) construction management philosophies.

Existing knowledge management perspectives and approaches are then presented and discussed, highlighting some of the existing knowledge creation frameworks and their application in the construction sector.

The nature and role of Knowledge Management Systems are discussed, including their different forms and functionality with examples drawn from the construction industry.

The book then discusses ways in which buildings are conceptualised and how these conceptualisations have been used to improve data, information, and knowledge management capabilities of individuals and teams within projects and organisations. An example of an ontology is presented, developed from a construction industry standard taxonomy (to provide the seeds of the ontology), enriched and expanded with additional concepts extracted from large discipline-oriented document bases using Information retrieval techniques.

The role of evolutionary algorithms in solving complex problems is then discussed, with an application using genetic algorithm in assisting designers in their architectural interventions.

The book then reflects on several decades of IT development in the construction industry, and attempts to map current efforts with a view of defining directions for the future of knowledge management from a technology perspective. The book argues that semantic process driven approaches with total lifecycle considerations should underpin ways in which we conceive, design, construct and maintain buildings.

The book then looks into organisational forms to support this vision. In particular, adoption and diffusion of knowledge-driven technologies involve addressing the specificities of small and medium-size enterprises that form a large proportion of the construction workforce. The book argues that SME alliance modes of operations promote business process innovation and allow SMEs to compete in new ways, get better reward for their work and gain greater financial strength, which in turn will give them the financial capability to move forward and develop their products and services.

The social, organisational and technology aspects of a construction alliance are then discussed. Important issues are explored that needs addressing in order to negotiate the necessary transition from a traditional to a knowledge-driven organisation that engages effectively in knowledge driven alliances.

It is vital that the construction sector migrates to a knowledge value creation culture where technology assets, human networks, social capital, intellectual capital and change management must be blended successfully to ensure effective knowledge value creation. A further understanding of the social and cultural features which influence knowledge value creation in the fragmented socio-cultural environment of the construction industry is needed.

Clearly, existing research acknowledges the pivotal and strategic role that human networks, social capital, technology and intellectual capital play in enhancing value creation in the construction sector. Organisations should give employees the opportunity to work in a team rather than work individually. Therefore, the organisation's knowledge values must be created through the network of relationships possessed by employees. Strong social relationships are critical factors to create more opportunities for team members to participate in problem solving and decision making, and offer a range of different skills, abilities, knowledge and experience to ensure that creative ideas are supported. A knowledge-based organisation needs all of its employees to share a culture that promotes the virtues of knowledge acquisition, sharing and value creation.

References

Akintoye, A. and McKellar, T. (1997). 'Electronic data interchange in the UK construction industry'. RICS Research Paper Series, 2(4).

Alavi, M. and Leidner, D. (2001). 'Review: Knowledge management and knowledge management systems: conceptual foundations and research issues'. *MIS Quarterly*, 25(1): 107–136.

Allen, V. (2003). *The Future of Knowledge Increasing Prosperity through Value Networks*. Woburn, MA: Butterworth-Heinemann.

Amor, R. W. (2004). 'Supporting standard data model mappings'. In *Proceedings of EC-PPM 2004*, Istanbul, Turkey, 8–10 September, pp. 35–40.

Anumba, C. J. and Evbuomwan, N. F. O. (1999). 'A taxonomy for communication facets in concurrent life-cycle design and construction'. *Computer-Aided Civil Infrastructure Engineering*, 14: 37–44.

Aranda, D. A. and Molina-Fernandez, L. M. (2002). 'Determinants of innovation through a knowledge-based theory lens'. *Industrial Management and Data Systems*, 102(5): 289–296.

Argyris, C. (1999). *On Organizational Learning*, 2nd edn. Oxford: Blackwell Business.

Artto, K. A. (1998). *Global Project Business and Dynamics of Change*. Helsinki: TEKES/PMA.

Ashby, W. R. (1956). *An Introduction to Cybernetics*. Methuen: London.

Aspin, R., DaDalto, L., Fernando, T., Gobbetti, E., Marache, M., Shelbourn, M. and Soubra, S. (2001). 'A conceptual framework for multi-modal interactive virtual workspace'. *ITcon*, 6, Special Issue: Information and Communication Technology Advances in the European Construction Industry: 149–162, http://www.itcon.org/2001/11.

Augenbroe, G. (1994). 'An overview of the COMBINE project'. In *First European Conference on Product and Process Modeling in the Building Industry*, Dresden, Germany, pp. 547–554.

Augenbroe, G. (1995). *COMBINE-2 Final Report*. CEC-DGXII: Brussels, Belgium.

Ayas, K. (1997). 'Design for learning for innovation: project management for new product development'. Thesis Publications, Erasmus University, Rotterdam.

Aziz, Z., Anumba, C. J., Ruikar, D., Carrillo, P. and Bouchlaghem, D. (2006). 'Intelligent wireless web services for construction: a review of enabling technologies'. *Automat Construct*, 15(2): 113–123.

Baeza-Yates, R. and Ribeiro-Neto, B. (1999). *Modern Information Retrieval*. New York: ACM Press.

Barresi, S., Rezgui, Y., Lima, C. and Meziane, F. (2005). 'Architecture to support semantic resources interoperability'. In IHIS '05: *Proceedings of the First International Work-*

shop on Interoperability of Heterogeneous Information Systems, 31 October–5 November, Bremen, Germany.

Barrett, P. and Sexton, M. (2006). 'Innovation in small, project-based construction firms'. *British Journal of Management*, 17: 331–346.

Bazjanac, V. (2004). 'Building energy performance simulation as part of interoperable software environments'. *Building and Environment*, 39(8): 879–883.

Becerra-Fernandez, I. and Sabherwal, R. (2001). 'Organizational knowledge management: a contingency perspective'. *Journal of Management Information Systems*, 18(1): 23–55.

Bechhofer, S., Horrocks, I., Goble, C. and Stevens, R. (2001). 'OilEd: a reason-able ontology editor for the Semantic Web'. In *Joint German/Austrian conference on Artificial Intelligence (KI01), Lecture Notes in Artificial Intelligence*, Vol. 2174. Springer: Berlin, pp. 396–408.

Beckhard, R. and Harris, R. (1987). *Transitions: Managing Complex Change*. London: Addison Wesley.

Bennett, J. and Jayes, S. (1995). *Trusting the Team: The Best Practice Guide to Partnering in Construction*. London: Thomas Telford Publishing.

Bennett, J. and Jayes, S. (1998). *The Seven Pillars of Partnering: A Guide to Second Generation Partnering*. London: Thomas Telford Publishing and Reading Construction Forum.

Bennis, W., Benne, K. and Chin, R. (1985). *The Planning of Change*, 4th edn. New York: Holt, Rinehart & Winston.

Bernaras, A., Laresgoiti, I. and Corera, J. (1996). 'Building and reusing ontologies for electrical network applications'. In *Proceedings of the European Conference on Artificial Intelligence (ECAI 96)*, Budapest, Hungary, pp. 298–302.

Björk, B.-C. (1989). 'Basic structure of a proposed building product model'. *Computer-Aided Design*, 21(2): 71–78.

Björk, B.-C. (1994). 'RATAS Project – developing an infrastructure for computer-integrated construction'. *Journal of Computing in Civil Engineering*, 8, pp. 400–419.

Björk, B.-C. and Penttilä, H. (1998). 'A scenario for the development and implementation of a building product model standard'. *Advances in Engineering Software*, 11(4): 176–187.

Black, C., Akintoyeb, A. and Fitzgerald, E. (2000). 'An analysis of success factors and benefits of partnering in construction'. *International Journal of Project Management*, 18: 423–434.

Blackler, F. (1995). 'Knowledge, knowledge work and organizations: an overview and interpretation'. *Organization Studies*, 16(6): 1021–1046.

Blair, M. M. and Wallman, S. M. H. (2001). *Unseen Wealth: Report of the Brookings Task Force on Intangibles*. Washington, DC: Brookings Institution Press.

Blazquez, M., Fernandez, M., Garcia-Pinar, J. M. and Gomez-Perez, A. (1998). 'Building ontologies at the knowledge level using the ontology design environment'. In *Proceeding of the Knowledge Acquisition Workshop (KAW98)*, Banff, Alberta, Canada, 18–23 April 1998.

Boddy, S., Rezgui, Y., Cooper, G. and Wetherill, M. (2007). 'Computer integrated construction: a review and proposals for future directions'. *Advances in Engineering Software*, 38: 677–687.

Bohms, M., Tolman, F. and Storer, G. (1994). 'ATLAS, A STEP towards computer integrated large scale engineering'. *Revue Internationale CFAO*, 9: 325–337.

Boisot, M. (1998). *Knowledge Assets: Securing Competitive Advantage in the Information Economy*. Oxford: Oxford University Press.

Bonaccorsi, A., Pammolli, F., Paoli, M. and Tani, S. (1999). 'Nature of innovation and technology management in system companies'. *R&D Management*, 29(1): 57–69.

Borghoff, U. M. and Pareschi, R. (1998). *Information Technology for Knowledge Management*. New York: Springer.

Bosch, K. O., Bingley, P. and van der Wolf, P. (1991). 'Design flow management in the NELSIS CAD framework'. In *Proceedings of the 28th Conference on ACM/IEEE Design Automation*, San Francisco, California, United States, June 17–22.

Bougrain, F. and Haudeville, B. (2002). 'Innovation, collaboration and SMEs internal research capacities'. *Research Policy*, 31(5): 735–747.

Bradshaw, J. (1996). 'Deriving strategies for reducing pollution damage using a genetic algorithm'. PhD thesis, Cardiff School of Engineering, Cardiff University.

Bradshaw, J. and Miles, J. C. (1997). 'Using standard fitness with genetic algorithms'. *Advances in Engineering Software*, 28: 425–435.

Broadbent, B. (2002). *ABCs of e-Learning: Reaping the Benefits and Avoiding the Pitfalls*. San Francisco, CA: Jossey-Bass/Pfeiffer.

Broglio, J., Callan, J. P., Croft, W. B. and Nachbar, D. W. (1995). 'Document retrieval and routing using the INQUERY system'. In Harman, D. K. (ed.), *Overview of the Third Retrieval Conference (TREC-3)*. NIST Special Publication 500–225, pp. 29–38.

Brown, J. and Duguid, P. (1998). 'Organizing knowledge'. *California Management Review*, 40(3): 90–112.

Brown, A., Rezgui, Y., Cooper, G., Yip, J. and Brandon, P. (1996). 'Promoting computer integrated construction through the use of distribution technology'. *ITcon*, 1: 51–67.

BS 6100 (1992). Glossary of building and civil engineering terms. British Standards Institution, http://bsonline.techindex.co.uk.

Burrell, G. and Morgan, G. (1979). *Sociological Paradigms and Organisational Analysis*. London: Heineman.

Carlsson, S. A., El Sawy, O. A., Eriksson, I. and Raven, A. (1996). 'Gaining competitive advantage through shared knowledge creation: in search of a new design theory for strategic information systems'. In *Proceedings of The Fourth European Conference on Information Systems*, Lisbon.

CERF (2000). 'Guidelines for moving innovations into practice'. Working Draft Guidelines for the CERF International Symposium and Innovative Technology Trade Show 2000, CERF, Washington, DC.

Chase, R. L. (1997). 'Knowledge management benchmarks'. *Journal of Knowledge Management*, 1: 83–92.

Checkland, P. (1981). *Systems Thinking, Systems Practice*. Chichester, UK: Wiley.

Checkland, P. and Holwell, S. (1998). *Information, Systems and Information Systems*. Chichester, UK: Wiley.

Checkland, P. and Scholes, J. (1990). *Soft Systems Methodology in Action*. Chichester, UK: Wiley.

Choo, C. W. (2000). 'Closing the cognitive gaps: how people process information'. In Marchand, D., Davenport, T. and Dickson, T. (eds), *Mastering Information Management*. Harlow: FT-Prentice-Hall, pp. 245–253.

Choo, C. W. (2003). 'Perspectives on managing knowledge in organizations'. *Cataloging and Classification Quarterly*, 37(2): 205–220.

Christensen, L. C., Christensen, T. R., Jin, Y., Kunz, J. and Levitt, R. E. (1997). 'Object oriented enterprise modeling and simulation of AEC projects'. *Computer-Aided Civil and Infrastructure Engineering*, 12(1): 157–170.

Cimiano, P. and Volker, J. (2005). 'Text2Onto: a framework for ontology learning and data-driven change discovery'. In *Proceedings of the 10th International Conference on Applications of Natural Language to Data Bases (NLDB'05)*, Alicante, Spain, pp. 227–238.

Clarke, A., Miles, J. C. and Rezgui, Y. (2009). 'Evolutionary algorithms for fire and rescue decision making'. In *FIRESEAT Conference*, University of Edinburgh, UK.

Cohen, S. G. and Bailey, D. E. (1997). 'What makes teams work: group effectiveness research from the shop floor to the executive suite'. *Journal of Management*, 23(3): 239–290.

Connaughton, S. L. and Daly, J. A. (2004). 'Identification with leader: a comparison of perceptions of identification among geographically dispersed and co-located teams'. *Corporate Communications: An International Journal*, 9(2): 89–103.

Cooper, G., Cerulli, C., Lawson, B. R., Peng, C. and Rezgui, Y. (2005). 'Tracking decision-making during architectural design'. *ITcon*, 10: 125–139, http://www.itcon.org/2005/10.

Corcho, O., Fernando-Lopez, M. and Gomez-Perez, A. (2003). 'Methodologies, tools and languages for building ontologies. Where is their meeting point?' *Data and Knowledge Engineering*, 46: 41–64.

Crampton, C. (2001). 'The mutual knowledge problem and its consequences for dispersed collaboration'. *Organization Science*, 12: 346–371.

Cristani, M. and Cuel, R. (2005). 'A survey on ontology creation methodologies'. *International Journal on Semantic Web and Information Systems*, 1(2): 49–69.

Curbera, F. (2002). 'Unravelling the web services web: an introduction to SOAP'. *WSDL, UDDI, IEEE Internet Computing*, 6(2): 86–93.

Dahlbom, B. and Mathiassen, L. (1995). *Computers in Context: The Philosophy and Practice of Systems Design*. Oxford: Blackwell Publishers.

Dainty, A. R. J., Briscoe, G. and Millett, S. J. (2001). 'Subcontractor perspectives on supply chain alliances'. *Construction Management and Economics*, 19(8): 841–848.

Dale, A. and Newman, L. (2008). 'Social capital: a necessary and sufficient condition for community development?' *Community Development Journal*, 45(1): 5–21.

Daniell, J. and Director, S. W. (1989). 'An object oriented approach to CAD tool control within a design framework'. In *Proceedings of the 26th ACM/IEEE Conference on Design Automation*, Las Vegas, Nevada, United States, June 25–28, 1989.

Davenport, T. and Prusak, L. (1998). *Working Knowledge: How Organizations Manage What They Know*. Boston, MA: Harvard Business School Press.

Davis, F. D. (1989). 'Perceived usefulness, perceived ease of use, and user acceptance of information technology'. *MIS Quarterly*, 13: 319–340.

Davis, F. D. (1993). 'User acceptance of information technology: system characteristics, user perceptions and behavioural impacts'. *International Journal of Man–Machine Studies*, 18(3): 475–487.

Deetz, S. (1992). *Democracy in an Age of Corporate Colonization: Developments in Communication and the Politics of Everyday Life*. Albany, NY: State University of New York Press.

Deetz, S. (1996). 'Describing differences in approaches to organization science: rethinking Burrell and Morgan and their legacy'. *Organization Science*, 7: 191–207.

DeFillippi, R. J. and Arthur, M. B. (1998). 'Paradox in project-based enterprise: the case of film making'. *California Management Review*, 40(2): 1–15.

Despres, C. and Chauvel, D. 1999. 'Knowledge management(s)'. *Journal of Knowledge Management*, 3(2): 110–123.

De Wilde, P., Rafiq, Y. and Packham, I. (2009). 'A study of thermal storage in buildings using interactive visualisation'. In Barjenbruch, M. *et al.* (eds), *Computation in Civil*

Engineering. Proceedings of the EG-ICE conference 2009, Technische Universität Berlin, pp. 67–74.

Ding, H. B. and Peters, L. S. (2000). 'Inter-firm knowledge management practices for technology and new product development in discontinuous innovation'. *International Journal of Technology Management*, 20(5–8): 588–600.

Dretske, F. (1981). *Knowledge and the Flow of Information*. Cambridge, MA: MIT Press.

Drucker, P. (1993). *Post-Capitalist Society*. Oxford: Butterworth-Heinemann.

Earl, M. (2001). 'Knowledge management strategies: toward a taxonomy'. *Journal of Management Information Systems*, 18(1): 215–233.

Eastman, C. M. (1992). 'Modelling of buildings: evolution and concepts'. *Automation in Construction*, 1: 99–109.

Eastman, C. M. (1999). *Building Product Models*. Boca Raton, FL: CRC Press.

Eastman, C. M. and Siabiris, A. (1995). 'A generic building product model incorporating building type information'. *Automation in Construction*, 3: 283–304.

Eastman, C., Wang, F., You, S.-J. and Yang, D. (2004). 'Deployment of an AEC industry sector product model'. *Computer-Aided Design*, 37(12): 1214–1228.

eCognos Consortium (2003). eCognos Field Trials, Project Deliverable D4.2, Derbi, France.

eConstruct (2001). Final edition of the bcXML Specifications, http://www.econstruct. org/6-public/bcxml_cd/publicdeliverables/d103_v2.pdf.

Eden, C. and Ackermann, F. (1998). *Making Strategy: The Journey of Strategic Management*. London: Sage.

Egan, J. (1998). *Rethinking Construction. The Report of the Construction Task Force*. London: HMSO, http://www.construction.detr.gov.uk/cis/rethink/.

Ekholm, A. (1996). 'A conceptual framework for classification of construction works'. *Journal of Information Technology in Construction*, 1: 25–50.

El-Diraby, T. A., Lima, C. and Fies, B. (2005). 'Domain taxonomy for construction concepts: toward a formal ontology for construction knowledge'. *Journal of Computing in Civil Engineering*, 19(4): 394–406.

Espejo, R. (1993). 'Giving requisite variety to strategy and information systems'. In Stowell, F. A. *et al.* (eds), *Systems Science*. New York: Plenum Press.

Espinosa, J. A., Cummings, J. N., Wilson, J. M. and Pearce, B. M. (2003). 'Team boundary issues across multiple global firms'. *Journal of Management Information Systems*, 19(4): 157–190.

European Commission (1996). *Green Paper on Innovation*, Bulletin of the European Union: Supplement 5/95. European Commission: Brussels.

Euzenat, J. (1996). 'Corporative memory through cooperative creation of knowledge bases and hyper-documents'. In *Proceedings of the 10th Knowledge Acquisition for Knowledge-Based Systems Workshop (KAW96)*, pp. 1–18.

Fahey, L. and Prusak, L. (1998). 'The eleven deadliest sins of knowledge management'. *California Management Review*, 40(3): 265–277.

Farquhar, A, Fikes, R. and Rice, J. (1996). 'The Ontolingua Server: a tool for collaborative ontology construction'. In *Proceedings of the 10th Knowledge Acquisition for Knowledge-Based Systems Workshop (KAW96)*, Banff, Alberta, pp. 44.1–44.19.

Fellows, R. and Liu, A. (2003). *Research Methods for Construction*. Oxford: Blackwell Publishing, p. 263.

Fernandez-Lopez, M., Gomez-Perez, A., Pazos-Sierra, A. and Pazos-Sierra, J. (1999). 'Building a chemical ontology using METHONTOLOGY and the ontology design environment'. *IEEE Intelligent Systems and Their Applications*, 4(1): 37–46.

Ferris, C. and Farrell, J. (2003). 'What are web services?' *Communications of the ACM*, 46(6): 31.

Firestone, J. M. and McElroy, M. W. (2003). *Key Issues in the New Knowledge Management*. Burlington, MA: Butterworth-Heinemann.

Fisher, K. and Fisher, M. D. (2001). *The Distance Manager: A Hands-On Guide to Managing Off-Site Employees and Virtual Teams*. New York: McGraw-Hill.

Franco, L. A., Cushman, M. and Rosenhead, J. (2004). 'Project review and learning in the construction industry: embedding a problem structuring method within a partnership context'. *European Journal of Operational Research*, 152(3): 586–601.

Friend, J. and Hickling, A. (1997). *Planning Under Pressure: the Strategic Choice Approach*. Oxford: Butterworth-Heinemann.

Furnas, G. W., Deerwester, S., Dumais, S. T., Landauer, T. K., Harshman, R. A., Streeter, L. A. and Lochbaum, K. E. (1988). 'Information retrieval using a singular value decomposition model of latent semantic structure'. In *Proceedings of the 11th Annual International Conference on Research and Development in Information Retrieval (ACM SIGIR)*, pp. 465–480.

Gabriel, Y. (2000). *Storytelling in Organisations*. Oxford: Oxford University Press.

Gann, D. and Salter, A. (2000). 'Innovation in project based, service-enhanced firms: the construction of complex products and systems'. *Research Policy*, 29(7–8): 955–972.

Gebert, H., Geib, M., Kolbe, L. and Brenner, W. (2003). 'Knowledge-enabled customer relationship management: integrating customer relationship management and knowledge management concepts'. *Journal of Knowledge Management*, 7(5): 107–123.

Gere, J. and Timoshenko, S. (1997). *Mechanics of Materials*, 4th edn. Boston, MA: PWS Publishing Company.

Giannini, F. Monti, M. Biondi, D., Bonfatti, F. and Daniela Monari, P. (2002). 'A modelling tool for the management of product data in a co-design environment'. *Computer-Aided Design*, 34(14): 1063–1073.

Gold, A. H., Malhotra, A. and Segars, A. H. (2001). 'Knowledge management: an organizational capabilities perspective'. *Journal of Management Information Systems*, 18(1): 185–214.

Goldberg, D. E. (1989). *Genetic Algorithms in Search, Optimisation and Machine Learning*. New York: Addison-Wesley.

Goleman, D. (1985). *Vital Lies, Simple Truths: The Psychology of Self-Deception*. London: Bloomsbury.

Gomez-Perez, A. (2001). 'Evaluation of ontologies'. *International Journal of Intelligent Systems*, 16(3): 10:1–10:36.

Goranson, H. T. (1999). *The Agile Virtual Enterprise: Cases, Metrics, Tools*. Westport, CT: Quorum Books.

Grant, R. M. (1997). *Contemporary Strategic Analysis: Concepts, Techniques, Applications*. Oxford: Blackwell Publishers.

Grierson, D. E. and Khajehpour, S. (2002). 'Method for conceptual design applied to office buildings'. *Journal of Computing in Civil Engineering*, 16(2): 83–103.

Griffiths, D. R. and Miles, J. C. (2003). 'Determining the optimal cross-section of beams'. *Advanced Engineering Informatics*, 17: 59–76.

Grilo, A., Jardim-Gonçalves, R., Steiger-Garcao, A. (2005). 'Shifting the construction interoperability paradigm, in the advent of Service oriented and model driven architectures'. In Scherer, R. J., Katranuschkov, P. and Schapke, S.-E. (eds), *Proceedings of CIB-W78 2005 22nd Conference on Information Technology in Construction*, July 2005.

Gruber, T. R. (1995). 'Toward principles for the design of ontologies used for knowledge sharing'. *International Journal of Human and Computer Studies*, 43: 907–928.

Grüninger, M. and Fox, M. S. (1995). 'Methodology for the design and evaluation of ontologies'. In *Workshop on Basic Ontological Issues in Knowledge Sharing*, Montreal.

Gu, P. and Chan, K. (1995). 'Product modelling using STEP'. *Computer-Aided Design*, 27(3): 163–179.

Guarino, N. and Welty, C. (2000). 'Supporting ontological analysis of taxonomic relationships'. In *Proceedings of the 19th International Conference on Conceptual Modeling (ER'00)*, Vol. 1920 of Lecture Notes in Computer Science. Berlin: Springer, pp. 210–224.

Gudivada, V., Raghavan, V., Grosky, W. and Kasanagottu, R. (1997). 'Information retrieval on the world wide web'. *IEEE Internet Computing* 1(5) (October–November): 58–68.

Hahn, U. and Schulz, S. (2003). 'Towards a broad-coverage biomedical ontology based on description logics'. *Pacific Symposium on Biocomputing*, 8: 577–588.

Halfawy, M. and Froese, T. (2005). 'Building integrated architecture/engineering/construction systems using smart objects: methodology and implementation'. *Journal of Computing in Civil Engineering*, 19(2): 172–181.

Hansen, M. T., Nohria, N. and Tierney, T. (1999). 'What's your strategy for managing knowledge?' *Harvard Business Review* (March–April): 101–116.

Hedlund, G. (1994). 'A model of knowledge management and the N-Form Corporation'. *Strategic Management Journal*, 15: 73–90.

Heinrichs, J. and Lim, J. (2005). 'Model for organizational knowledge creation and strategic use of information'. *Journal of the American Society for Information Science and Technology*, 56(6): 620–629.

Hjelt, M. and Bjork, B.-C. (2006). 'Experiences of edm usage in construction projects'. *Journal of Information Technology in Construction*, 11:113–125.

Holland, J. H. (1975). *Adaptation in Natural and Artificial Systems*. Ann Arbor: University of Michigan Press.

Holsapple, C. W and Joshi, K. D. (2002). 'A collaborative approach to ontology design'. *Communications of the ACM*, 45(2): 42–47.

Honey, P. (1998). 'The debate starts here (a declaration on learning)'. In *The e-Learning Revolution: From Propositions to Action*. London: Chartered Institute of Personnel and Development, pp. 179–196.

Hooper, J. N. (1993). 'A knowledge-based system for strategic sludge disposal planning'. PhD thesis, Cardiff School of Engineering, Cardiff University.

Howard, N., Bennett, P., Bryant, J.W. and Bradley, M. (1993). 'Manifesto for a theory of drama and irrational choice'. *Journal of the Operational Research Society*, 44(1): 99–103.

Howard, R. and Bjork, B.-C. (2008). 'Building information modelling: experts' views on standardisation and industry deployment'. *Advanced Engineering Informatics*, 22: 271–280.

Huczynski, A. and Buchanan, D. (2001). *Organizational Behaviour: An Introductory Text*, 4th edn. London: FT/Prentice Hall.

Huseby, T. and Chou, S. T. (2003). 'Applying a knowledge-focused management philosophy to immature economies'. *Industrial Management and Data Systems*, 102(1): 17–25.

Huysman, M. and de Wit, D. (2002). *Knowledge Sharing in Practice*. Dordrecht: Kluwer Academics.

Huysman, M. and Wulf, V. (2006). 'IT to support knowledge sharing in communities, towards a social capital analysis'. *Journal of Information Technology*, 21: 40–51.

IFC (2010). Building smart. In 'Building Smart Web Site', 2010. http://www.buildings-mart.com.

ISO 9735:1988 (1988). Electronic Data Interchange for Administration, Commerce and Transport (EDIFACT). International Standards Organization. TC 154 (http://www.iso.ch/cate/d17592.html).

ISO 12006–2 (2001). Building construction: organization of information about construction works (Part 2: framework for classification of information). International Standards Organisation.

Jarvenpaa, S. and Leidner, D. (1999). 'Communication and trust in global virtual teams'. *Organization Science*, 10(6): 791–815.

Jeng, T. S. and Eastman, C. M. (1999). 'Design process management'. *Computer-Aided Civil and Infrastructure Engineering*, 14(1): 55–67.

Jongeling, R. Emborg, M. and Olofsson, T. (2005). 'nD modelling in the development of cast in place concrete structures', *ITcon*, 10, Special Issue: From 3D to nD modelling: 27–41 (http://www.itcon.org/2005/4)

Kalay, Y. E. (2001). 'Enhancing multi-disciplinary collaboration through semantically rich representation'. *Automation in Construction*, 10(6): 741–755.

Kalyanpur, A., Parsia, B., Sirin, E. and Hendler, J. (2005). 'Debugging unsatisfiable classes in OWL ontologies'. *Journal of Web Semantics*, 3(4): 268–293.

Kanter, R. M. (1983). *The Change Masters*. New York: Simon & Schuster.

Kanter, R. M. (1997). *The Frontiers of Management*. Boston, MA: Harvard Business School Press.

Katranuschkov, P., Scherer, R. and Turk, Z. (2001). 'Intelligent services and tools for concurrent engineering? An approach towards the next generation of collaboration platforms'. *ITcon*, 6, Special Issue: Information and Communication Technology Advances in the European Construction Industry: 111–128 (http://www.itcon.org/2001/9).

Katranuschkov, P., Gehre, A. and Scherer, R. J. (2003). 'An ontology framework to access IFC model data'. *ITcon*, 8: 413–437, Special Issue: eWork and eBusiness. http://www.itcon.org/2003/29.

Kayworth, T. and Leidner, D. (2000). 'The global virtual manager: a prescription for success'. *European Management Journal*, 18(2): 183–194.

Kean, A. J. (2004). 'Design search and optimisation using radial basis functions with regression capabilities'. In Parmee, I. C. (ed.), *Adaptive Computing in Design and Manufacture VI*. London: Springer, pp. 39–50.

Kelly, G. A. (1955). *The Psychology of Personal Constructs*. New York: W.W. Norton.

Khajehpour, S. and Grierson, D. E. (1999). 'Filtering of pareto-optimal trade-off surfaces for building conceptual design'. In Topping, B. H. V. and Kumar, B. (eds), *Optimzation and Control in Civil and Structural Engineering*. Edinburgh, UK: Civil-Comp Press, pp. 63–70.

Khajehpour, S. and Grierson, D. E. (2003). 'Profitability versus safety of high-rise office buildings'. *Journal of Structural and Multidisciplinary Optimisation*, 25: 1–15.

Kietz, J. U., Maedche, A. and Volz, R. (2000). 'A method for semi-automatic ontology acquisition from a corporate intranet'. In *Proceedings of the International Conference on Knowledge Engineering and Knowledge Management (EKAW'00)*, Workshop on Ontologies and Texts, CEUR Proceedings, Juan-Les-Pins.

Kim, J., Pratt, M. J, Iyer, R. G. and Sriram, R. D. (2008). 'Standardized data exchange of CAD models with design intent'. *Computer-Aided Design*, 40(7): 760–777.

Klein, M. (2001). 'Combining and relating ontologies: an analysis of problems and solutions'. In Gomez-Perez, A., Gruninger, M., Stuckenschmidt, H. and Uschold, M. (eds), *Workshop on Ontologies and Information Sharing. International Joint Conferences on Artificial Intelligence (IJCAI'01)*, Seattle, USA, 4–5 August.

Klein, M., Fensel, D., Kiryakov, A. and Ognyanov, D. (2002). 'Ontology versioning and change detection on the Web'. In Gomez-Perez, A. and Benjamins, V. R. (eds), *13th International Conference on Knowledge Engineering and Knowledge Management (EKAW02)*, Lecture Notes in Artificial Intelligence, Vol. 2473. Berlin: Springer.

Koenig, M. E. D. (2002). 'The third stage of KM emerges'. *KMWorld*, 11(3): 20–21.

Koo, B. and Fischer, M. (2000). 'Feasibility study of 4D CAD in commercial construction'. *Journal of Construction Engineering and Management*, 126(4): 251–260.

Larsson, R., Bengtsson, L., Henriksson, K. and Sparks, J. (1998). 'The interorganizational learning dilemma: collective knowledge development in strategic alliances'. *Organization Science*, 9(3): 285–305.

Latham, M. (1994). *Constructing the Team: Final Report of the Government/Industry Review of Procurement and Contractual Arrangements in the UK Construction Industry*. London: HMSO.

Leavitt, H. (1965). 'Applied organizational change in industry: structural, technological and humanistic approaches'. In March, J. (ed.), *Handbook of Organizations*. Chicago, IL: Rand McNally, pp. 1144–1170.

Lee, A., Wu, S., Aouad, G. and Fu, C. (2002) 'Towards nD modelling'. In *Proceedings of the European Conference on Information and Communication Technology Advances and Innovation in the Knowledge Society (E-sm@art)*, Salford.

Lee, G. and Cole, R. (2003). 'From a firm-based to a community-based model of knowledge creation'. *Organization Science*,14(6): 633–649.

Lee, J. (2001). 'The impact of knowledge sharing, organizational capability and partnership quality on IS outsourcing success'. *Information and Management*, 38: 323–335.

Lesser, E. L. (2000). *Knowledge and Social Capital: Foundations and Applications*. Boston, MA: Butterworth Heinemann.

Levy, M., Loebbecke, C. and Powell, P. (2001). 'SMEs, co-opitition and knowledge sharing: the IS role'. In *Global Co-Operation in the New Millennium. The 9th European Conference on Information Systems*, Bled, Slovenia.

Liebowitz, J. and Suen, C. Y. (2000). 'Developing knowledge management metrics for measuring intellectual capital'. *Journal of Intellectual Capital*, 1: 54–67.

Lima, C., El-Diraby, T. and Stephens, J. (2005). 'Ontology-based optimisation of knowledge management in e-construction'. *ITcon*, 10: 305–327.

Lipnack, J. and Stamps, J. (2000). *Virtual Teams: People Working Across Boundaries with Technology*, 2nd edn. New York: Wiley & Sons.

Lowe, R. (2001). *A Review of Recent and Current Initiatives on Climate Change and its Impact on the Built Environment: Impact, Effectiveness and Recommendations*. Centre for the Built Environment, Leeds Metropolitan University.

Löwnertz, K. (1998). 'Change and exchange : electronic document management in building design'. Licentiate Thesis, Dept. of Construction Management and Organisation, Royal Institute of Technology, Stockholm, Sweden.

Lyytinen, K. (1987). 'A taxonomic perspective of information systems development: theoretical constructs and recommendations'. In Boland, R. J. and Hirschheim, R. A. (eds), *Critical Issues in Information Systems Research*, Wiley Series In Information Systems. New York: John Wiley & Sons, pp. 3–41.

McAdam, R. and Galloway, A. (2005). 'Enterprise resource planning and organisational innovation: a management perspective'. *Industrial Management and Data Systems*, 105(3): 280–290.

McAdam, R. and McCreedy, S. (1999). 'A critical review of knowledge management models'. *The Learning Organisation*, 6(3): 91–100.

McDermott, R. and O'Dell, C. (2001). 'Overcoming cultural barriers to sharing knowledge'. *Journal of Knowledge Management*, 5(1): 1367–3270.

McDonough, E., Kahn, K. and Barczak, G. (2001). 'An investigation of the use of global, virtual, and collocated new product development teams'. *Journal of Product Innovation Management*, 18(2): 110–120.

McElroy, M. W. (1999). 'The second generation of knowledge management', *Knowledge Management* (October): 86–88.

McGuiness, D. L., Fikes, R., Rice, J. and Wilder, S. (2000). 'An environment for merging and testing large ontologies'. In *Seventh International Conference on Principles of Knowledge Representation and Reasoning 5KR 2000*, Breckenridge, Colorado, USA: 12–15.

Machlup, F. and Mansfield, U. (1983). *The Study of Information: Interdisciplinary Messages*. New York: Wiley & Sons.

McQueen, R. (1998). 'Four views of knowledge and knowledge management'. In *Proceedings of the Fourth Americas Conference on Information Systems*.

Mannisto, T., Peltonen, H., Martio, A. and Sulonen, R. (1998). 'Modelling generic product structures in STEP'. *Computer-Aided Design*, 30(14): 1111–1118.

Marir, F., Rezgui, Y. and Benhadj, R. (2000). 'A case-based expert system for construction project process activity specifications'. *International Journal in Construction Information Technology*, 8(1): 53–73.

Markus, L. M. (2001). 'Toward a theory of knowledge reuse: types of knowledge reuse situations and factors in reuse success'. *Journal of Management Information Systems*,18(1): 57–93.

Marshall, A. (1965). *Principles of Economics*. London: Macmillan.

Maznevski, M. and Chudoba, K. (2001). 'Bridging space over time: global virtual team dynamics and effectiveness'. *Organization Science*, 11: 473–492.

Miles, J. C., Kwan, A, Wang, K. and Zhang, Y. (2007). 'Searching for good topological solutions using evolutionary algorithms'. In Topping, B. (ed), *Civil Engineering Computations: Tools and Techniques*. Stirling, Scotland: Saxe-Coburg Publications, pp. 149–172.

NCCTP. (2006). 'Proving collaboration pays'. Available at: http://ncctp.constructingexcellence.org.uk/downloads/making_collaboration_pay.pdf (accessed 28/02/10).

Neef, D. (1999). 'Making the case for knowledge management: the bigger picture'. *Management Decision*, 37(1): 72–78.

NIBS. (2007). National Institute of Building Sciences. BIM Committee Web site at http://www.nibs.org/BIMcommittee.html (accessed 13/07/07).

Nititthamyong, P. and Skibniewski, M. (2006). 'Success/failure factors and performance measures of web-based construction project management systems: professionals' viewpoint'. *Journal of Construction Engineering and Management*, 132(1): 80–87.

Nonaka, I. and Konno, N. (1998). 'The concept of "Ba": building a foundation for knowledge creation', *California Management Review*, 40(3): 40–54.

Nonaka, I. and Takeuchi, H. (1995). *The Knowledge-creating Company: How Japanese Companies Create the Dynamics of Innovation*. New York: Oxford University Press.

Nonaka, I., Toyama, R. and Konno, N. (2000). 'SECI, Ba and leadership: a unified model of dynamic knowledge creation'. *Long Range Planning*, 33: 5–34.

Noy, N. F. and Klein, M. (2002). *Ontology Evolution: Not the Same as Schema Evolution*, Technical Report SMI-2002–0926, Stanford Medical Informatics, CA.

Noy, N. F. and Musen, M. A. (2000). 'PROMPT: Algorithm and tool for automated ontology merging and alignment'. In *Proceedings of the 17th National Conference on Artificial Intelligence (AAAI'00)*, Austin, Texas.

Noy, N. F., Sintek, M., Decker, S., Crubézy, M., Fergerson, R. W. and Musen, M. A. (2001). 'Creating semantic web contents with Protégé-2000'. *IEEE Intelligent Systems*, 16(2): 60–71.

O'Brien, M. and Al-Soufi, A. (1994). 'A survey of data communications in the construction industry'. *Construction Management and Economics*, 12(5): 457–465.

Ogawa, Y., Morita, T. Kobayashi, K. (1991). 'A fuzzy document retrieval system using the key word connection matrix and a learning method'. *Fuzzy Sets and Systems*, (39): 163–179.

Oinas-Kukkonen, H. (2004). 'The 7C model for organisational knowledge creation and management'. In *Proceedings of The Fifth European Conference on Organizational Knowledge, Learning and Capabilities*, Innsbruck.

Omelayenko, B. (2001). 'Syntactic-level ontology integration rules for e-commerce'. In *Proceedings of the 14th FLAIRS Conference (FLAIRS-2001)*, Key West, Florida, May 21–23, 2001. Key West, FL: AAAI Press.

Orlikowski, W. and Baroudi, J. (1991) 'Studying information technology in organizations: research approaches and assumptions'. *Information Systems Research*, 2(1): 1–28.

O'Rourke, J. (1998). *Computational Geometry in C*, 2nd edn. Cambridge: Cambridge University Press.

OWL-S Coalition. (2005). OWL-S Coalition web site at http://www.daml.org/services/owl-s/1.0/ (accessed 15/07/10).

Pahl, G. and Beitz, W. (1988). *Engineering Design: A Systematic Approach*, 2nd edn, (translated into English by Pomerans, A. and Wallace, K.). London: The Design Council.

Pan, S. and Scarbrough, H. (1998). 'A socio-technical view of knowledge sharing at Buckman Laboratories'. *Journal of Knowledge Management*, 2: 55–66.

Parmee, I. C. (1998). 'Evolutionary and adaptive strategies for efficient search across whole system engineering design hierarchies'. *AIEDAM*: 431–445.

Parmee, I. C. (2001). *Evolutionary and Adaptive Computing in Engineering Design*. London: Springer-Verlag.

Partridge, C. (2002). *The Role of Ontology in Integrating Semantically Heterogeneous Databases*, Technical Report 05/02, LADSEB-CNR, Padova, Italy, June 2002. http://www.loa-cnr.it/Papers/ladseb_tr05-02.pdf.

Peters, T. and Waterman, R. H. (1982). *In Search of Excellence*. New York: Harper & Row.

Polanyi, M. (1966). *The Tacit Dimension*, New York: Doubleday.

Powell, A., Piccoli, G. and Ives, B. (2004). 'Virtual teams: a review of current literature and directions for future research'. *Database for Advances in Information Systems*, 35(1): 6–36.

Preiss, K., Goldman, S. L. and Nagel, R. N. (1996). *Cooperate to Compete. Building Agile Business Relationships*. New York: Van Nostrand Reinhold.

Raphael, B. and Smith, I. F. C (2003). *Fundamentals of Computer-Aided Engineering*, Chichester, UK: Wiley.

Rastogi, P. N. (2000). 'Knowledge management and intellectual capital: the new virtuous reality of competitiveness'. *Human Systems Management*, 19(1): 39–48.

Reddy, S.B. and Reddy, R. (2002). 'Competitive agility and the challenge of legacy information systems'. *Industrial Management and Data Systems*, 102(1): 5–16.

Rezgui, Y. (2001). 'Review of information and the state of the art of knowledge management practices in the construction industry'. *The Knowledge Engineering Review*, 16(3): 241–254.

Rezgui, Y. (2006). 'Ontology driven knowledge management using information retrieval techniques'. Computing in Civil Engineering, 20(3): 261–270.

Rezgui, Y. (2007a). 'Exploring virtual team-working effectiveness in the construction sector'. *Interacting with Computers*, 19(1): 96–112.

Rezgui, Y. (2007b). 'Knowledge systems and value creation: an action research investigation'. *Industrial Management and Data Systems*, 107(2): 166–182.

Rezgui, Y. (2007c). 'Role-based service-oriented implementation of a virtual enterprise: a case study in the construction sector'. *Computers in Industry*, 58 (1): 74–86.

Rezgui, Y. (2007d). 'Text-based domain ontology building using Tf-Idf and metric clusters techniques'. *The Knowledge Engineering Review*, 22: 379–403.

Rezgui, Y. and Cooper, G. (1998). 'A proposed open infrastructure for construction project document sharing'. *ITcon*, 3: 11–25.

Rezgui, Y. and Karstila, K. (1998). 'D120: Building process and document lifecycle analysis'. Condor ESPRIT 23105 D120 Deliverable. University of Salford.

Rezgui, Y. and Miles, J. (2009). 'Transforming SME strategies via innovative transient knowledge-based alliances in the construction sector'. In *Proceedings of the IEEE INDIN Conference on Industrial Informatics*, Cardiff, 23–26 June 2009, pp. 859–864.

Rezgui, Y. and Miles, J. C. (2010). 'Exploring the potential of SME alliances in the construction sector'. *Construction Engineering and Management* (Journal of the American Society of Civil Engineering), 136(5): 558–567. 10.1061/(ASCE)CO.1943-7862.0000150.

Rezgui, Y. and Nefti-Meziani, S. (2007). 'Ontology-based dynamic composition of services using semantic relatedness and categorisation techniques'. In *ICEIS, 9th International Conference on Enterprise Information Systems*, Funchal, Madeira, Portugal.

Rezgui, Y. and Zarli, A. (2006). 'Paving the way to digital construction: a strategic roadmap'. *Construction Engineering and Management*, 132(12): 767–776.

Rezgui, Y., Cooper, G. and Brandon, P. (1998). 'Information management in a collaborative multi-actor environment'. *Computing in Civil Engineering*, 12(3): 136–144.

Rezgui, Y., Wilson, I., Olphert, W. and Damodaran, L. (2005). 'Socio-organizational issues'. In Camarinha-Matos, L. M., Afsarmanesh, H. and Ollus, M. (eds), *Virtual Organizations Systems and Practices*. New York: Springer Science, pp. 177–198.

Rezgui, Y., Boddy, S., Wetherill, M. and Cooper, G. (2009). 'Past, present and future information and knowledge sharing in the construction industry: towards semantic service-based e-construction?' *Computer-Aided Design*, 10.1016/j.cad.2009.06.005.

Rezgui, Y., Hopfe, C. J. and Vorakulpipat, C. (2010). 'Generations of knowledge management in the architecture, engineering and construction industry: an evolutionary perspective'. *Advanced Engineering Informatics*, 24 (2): 219–228, 10.1016/j.aei.2009.12.001.

Richardson, J. T., Palmer, M. R., Liepins, G. and Hilliard, M. (1989). 'Some guidelines for genetic algorithms with penalty functions'. In Schaffer, J. D. (ed.), *Proceedings of the 3rd International Conference on Genetic Algorithms*. Los Altos, CA: Morgan Kaufmann, pp. 191–197.

Richter, S., Huhnt, W. and Wotschke, P. (2009). 'Applying a new approach for generating construction schedules for real projects'. In Barjenbruch, M. *et al.* (eds), *Computation in Civil Engineering, Proceedings of the EG-ICE conference 2009*, Technische Universitat Berlin, pp. 234–241.

Rijsbergen, C. J. (1979). *Information Retrieval*. London: Butterworths.

Roberts, J. (2000). 'From know-how to show-how? Questioning the role of information and communication technologies in knowledge transfer'. *Technology Analysis and Strategic Management*, 12(4): 429–443.

Robertson, S. and Sparck Jones, K. (1976). 'Relevance weighting of search terms'. *Journal of the American Society of Information Sciences*, 27(3): 129–145.

Rogers, E. M. (1995). *Diffusion of Innovations*, 4th ed. New York: Free Press.

Rosenhead, J. (2001). 'Robustness analysis: keeping your options open'. In Mingers, J. (ed.), *Rational Analysis for a Problematic World Revisited: Problem Structuring Methods for Complexity Uncertainty and Conflict*. Chichester, UK: Wiley, pp. 181–207.

Rosenhead, J. and Mingers, J. (eds) (2001). *Rational Analysis for a Problematic World Revisited: Problem Structuring Methods for Complexity, Uncertainty and Conflict*. Chichester, UK: Wiley.

Routledge (2000). *Concise Routledge Encyclopedia of Philosophy*. New York: Routledge.

Ruikar, K. and Emmitt, S. (2009). 'Editorial: Technology strategies for collaborative working'. *ITcon*, 14, Special Issue: Technology Strategies for Collaborative Working): 14–16 (http://www.itcon.org/2009/03).

SABLE. (2006). SABLE web site, http://www.blis-project.org/~sable/ (accessed 24/04/06).

Salton, G. and Buckley, C. (1988). 'Term weighting approaches in automatic retrieval'. *Information Processing and Management*, 24(5): 513–523.

Salton, G. and Lesk, M. (1968). 'Computer evaluation of indexing and text processing'. *Journal of the ACM*, 15(1): 8–36.

Salton, G. and Yang, C. (1973). 'On the specification of term values in automatic indexing'. *Journal of Documentation*, 29: 351–372.

Salton, G., Fox, E. and Wu, H. (1983). 'Extended Boolean information retrieval'. *CACM*, 26(11): 1022–1036.

Sarker, S., Lau, F. and Sahay, S. (2001). 'Using an adapted grounded theory approach for inductive theory building about virtual team development'. *Database for Advances in Information Systems*, 32: 38–56.

Schlenoff, C., Gruninger, M., Ciocoiu, M. and Lee, J. (1999) 'The essence of the process specification language'. Special Issue on Modeling and Simulation of Manufacturing Systems in *Transactions of the Society for Computer Simulation International*.

Schön, D. (1983). *The Reflective Practitioner: How Professionals Think in Action*. New York: Basic Books.

Schubert, P., Lincke, D. and Schmid, B. (1998). 'A global knowledge medium as a virtual community: the NetAcademy concept'. In *Proceedings of The Fouth Americas Conference on Information Systems*, Baltimore, Maryland.

Schultze, U. (1998). 'Investigating the contradictions in knowledge management'. In *Proceedings of IFIP Conference on Information Systems*, Helsinki.

Schultze, U. and Leidner, D. E. (2002). 'Studying knowledge management in information systems research: discourses and theoretical assumption'. *MIS Quarterly*, 26(3): 213–242.

Scott, C. R. and Fontenot, J. (1999). 'Multiple identifications during team meetings: a comparison of conventional and computer-supported interactions'. *Communication Reports*, 12: 91–100.

Senge, P. (1990). *The Fifth Discipline: The Art and Practice of the Learning Organization*. New York: Doubleday Currency.

Shadbolt, N. and Milton, N. (1999). 'From knowledge engineering to knowledge management'. *British Journal of Management*, 10: 309–322.

Shamos, M. I. (1978). 'Computational geometry'. PhD thesis, Yale University, New Haven, UMI #7819047.

Shaw, D. J., Miles, J. C. and Gray, W. A. (2005a). 'Conceptual design of orthogonal commercial buildings'. In Topping, B. (ed.), *Proceedings of the 8th International Conference on the Application of Artificial Intelligence to Civil, Structural and Environmental Engineering*. Civil-Comp Press: Stirling, UK, paper 1.

Shaw, D. J., Miles, J. C. & Gray, W. A. (2005b). 'Conceptual design of geodesic domes'. In Topping, B. (ed.), *Proceedings of the 8th International Conference on the Application of Artificial Intelligence to Civil, Structural and Environmental Engineering*. Civil-Comp Press: Stirling, UK, paper 28.

Shea, K. and Cagan, J. (1997). 'Innovative dome design: applying geodesic patterns with shape annealing'. *Artificial Intelligence for Engineering Design, Analysis and Manufacturing*, 11: 379–394.

Sieloff, C. G. (1999). '"If only HP knew what HP knows": the roots of knowledge management at Hewlett-Packard'. *Journal of Knowledge Management*, 3(1): 47–53.

Sisk, G. M. (1999). 'The use of a GA-Based DSS for realistically constrained building design'. PhD thesis, Cardiff School of Engineering, Cardiff University, 250pp.

Sloman, M. (2001). *The e-Learning Revolution: From Propositions to Action*. London: Chartered Institute of Personnel and Development.

Snowden, D. (2002). 'Complex acts of knowing: paradox and descriptive self-awareness'. *Journal of Knowledge Management*, 6(2): 1–14.

Solanki, M., Cau, A. and Zedan, H. (2004). 'Augmenting semantic Web service descriptions with compositional specification'. In *Proceedings of the 13th international conference on World Wide Web*, New York: ACM Press, pp. 544–552.

Soley R. (2000). 'Model driven architecture'. Object Management Group white paper, ftp://ftp.omg.org/pub/docs/omg/00-11-05.pdf (accessed 09/11/10).

Sor, R. (2004). 'Information technology and organisational structure: vindicating theories from the past'. *Management Decision*, 42(2): 316–329.

Staab, S., Schnurr, H. P., Studer, R. and Sure, Y. (2001). 'Knowledge processes and ontologies'. *IEEE Intelligent Systems*, 16(1): 26–34.

STEP – ISO 10303–1 (1994). 'Industrial automation systems and integration – Product data representation and exchange – Part 1: overview and fundamental principles. International Standards Organization 1994. TC 184/SC 4'. http://www.iso.ch/cate/d20579.html.

Studer, R., Benjamins, V. R. and Fensel, D. (1998). 'Knowledge engineering: principles and methods'. *Data and Knowledge Engineering*, 25: 161–197.

Stumme, G. and Maedche, A. (2001). 'Ontology merging for federated ontologies on the semantic web'. In Gomez-Perez, A., Gruninger, M., Stuckenschmidt, H. and Uschold, M. (eds), *Workshop on Ontologies and Information Sharing, International Joint Conferences on Artificial Intelligence (IJCAI'01)*, Seattle, Washington, 4–5 August 2001.

Suchan, J. and Hayzak, G. (2001). 'The communication characteristics of virtual teams: a case study'. *IEEE Transactions on Professional Communications*, 44: 174–186.

Sure, Y., Erdmann, M., Angele, J., Staab, S., Studer, R. and Wenke, D. (2002). 'OntoEdit: collaborative ontology engineering for the semantic web'. In *First International Semantic Web Conference (ISWC'02)*, Lecture Notes in Computer Science, Vol. 2342. Berlin: Springer, pp. 221–235.

Sveiby, K.-E. (1997). *The New Organizational Wealth: Managing and Measuring Knowledge-based Assets*. San Francisco: Berrett Koehler.

Sveiby, K. (1999). 'Industry led sharing of knowledge – The future of hi-tech education?', http://www.sveiby.com.

Swartout, B., Ramesh, P., Knight, K. and Russ, T. (1997). 'Toward distributed use of large-scale ontologies'. In *AAAI Symposium on Ontological Engineering*, Stanford, California.

Tan, B., Wei, K., Huang, W. and Ng, G. (2000). 'A dialogue technique to enhance electronic communication in virtual teams'. *IEEE Transactions on Professional Communications*, 43: 153–165.

Tapscott, D. (1996). *Digital Economy*. New York: McGraw-Hill.

Teece, D. (1998). 'Capturing value from knowledge assets: the new economy, markets for know-how, and intangible assets'. *California Management Review*, 40(3): 55–79.

Thomas, G. and Bone, R. (2000). *Innovation at the Cutting Edge: The Experience of Three Infrastructure Projects*, CIRIA Funders Report FR/CP/79, CIRIA, London.

Tiwana, A. (2000). *The Knowledge Management Toolkit: Practical Techniques for Building a Knowledge Management System*. Upper Saddle River, NJ: Prentice Hall.

Tomovic, A., Janicic, P. and Keselj, V. (2006). 'n-Gram-based classification and unsupervised hierarchical clustering of genome sequences'. *Computer Methods and Programs in Biomedicine*, 81: 137–153.

Tsai, W. and Ghoshal, S. (1998). 'Social capital and value creation: the role of intrafirm networks'. *Academy of Management Journal*, 41(4): 464–476.

Tuomi, I. (1999). 'Data is more than knowledge: implications of the reversed knowledge hierarchy for knowledge management and organizational memory'. *Journal of Management Information Systems*, 16(3): 103–117.

Turk, Z., Stankovski, V., Dolenc, M. and Cerovšek, T. (2004). 'Semantic grid infastructure for AEC virtual enterprise: a research agenda'. In *Proceedings of the International Conference on Construction Information Technology (INCITE 2004): World IT for Design and Construction*, Langkawi, Malaysia, 18–21 February.

Turner, J. (1988). 'AEC building systems model', working paper, O/TC/184/SC4/WG1.

Turtle, H. and Croft, W. B. (1990). 'Inference networks for document retrieval'. In *Proceedings of the 13th Annual International Conference ACM SIGIR Conference on Research and Development in Information Retrieval*, pp. 1–24.

Turtle, H. and Croft, W. B. (1991). 'Evaluation of an inference network-based retrieval model'. *ACM Transactions on Information Systems*, 9(3): 187–222.

Uschold, M. and King, M. (1995). 'Towards a methodology for building ontologies'. In the *Proceedings of the International Joint Conferences on Artificial Intelligence (IJCAI95), Workshop on Basic Ontological Issues in Knowledge Sharing*, Montreal.

Vakola, M. and Wilson, I. (2004). 'The challenge of virtual organization: critical success factors in dealing with constant change'. *Team Performance Management*, 10(5/6): 112–120.

Van Den Heuvel, W. J. and Maamar, Z. (2003). 'Moving towards a framework to compose intelligent web services'. *Communications of the ACM*, 46(10): 103–109.

Van Ryssen, S. and Hayes Godar, S. (2000). 'Going international without going international: multinational virtual teams'. *Journal of International Management*, 6: 49–60.

Vance, D. (1997). 'Information, knowledge and wisdom: the epistemic hierarchy and computer-based information systems'. In *AMCIS 1997 Proceedings*.

Venters, W. (2001). 'Literature review for C-Sand: Knowledge Management, C-SandD/DOC/1002/2'. London School of Economics.

Verhoeff, J., Goffmann, W. and Belzer, J. (1961). 'Inefficiency of the use of Boolean functions for information retrieval systems'. *CACM*, 4(12): 557–558.

Vorakulpipat, C. and Rezgui, Y. (2008). 'Value creation: the future of knowledge management'. *Knowledge Engineering Review*, 23(4): 283–294.

Warkentin, M. and Beranek, P. M. (1999). 'Training to improve virtual team communication'. *Information Systems Journal*, 9: 271–289.

W3C. (2001). 'Web services description language'. W3C Note, 15 March. W3C website at http://www.w3.org/TR/wsdl (accessed 01/06/09).

Wartick, S. (1992). 'Boolean operations'. In Frakes, W. B. and Baeza-Yates, R. A. (eds), *Information Retrieval: Data Strutures and Algorithms*. Englewood Cliffs, NJ: Prentice Hall, pp. 264–292.

Wasko, M. and Faraj, S. (2005). 'Why should I share? Examining social capital and knowledge contribution in electronic networks of practice'. *MIS Quarterly*, 29(1): 35–58.

Watson, I. and Marir, F. (1994). 'Case-based reasoning: a review'. *The Knowledge Engineering Review*, 9(4): 327–354.

Webber, A. (1984). 'What's so new about the new economy?' *Harvard Business Review* (January–February): 24–42.

Weick, K. (1995). *Sensemaking in Organisations*. London: Sage Publications.

Weick, K. (1996). 'Prepare your organisation to fight fires'. *Harvard Business Review* (May–June): 143–148.

Welty, C. and Guarino, N. (2001). 'Supporting ontological analysis of taxonomic relationships'. *Data and Knowledge Engineering*, 39(1): 51–74.

Wenger, E., McDermott, R. and Snyder, W. M. (2002). *Cultivating Communities of Practice: A Guide to Managing Knowledge*. Cambridge, MA: Harvard Business School Press.

Wetherill, M., Rezgui, Y., Lima, C. and Zarli. A. (2002). 'Knowledge management for the construction industry: the eCognos project'. *ITcon* (7): 183–196, Special Issue: ICT for Knowledge Management in Construction.

Wetherill, M., Rezgui, Y., Boddy, S. and Cooper, G. (2007). 'Intra and inter-organisational knowledge services to promote informed sustainability practices'. *Computing in Civil Engineering*, 21: 78–79.

Wheatcroft, J. (2000). 'Organizational change, the story so far'. *Industrial Management and Data Systems*, 100(1): 5–9.

Wiesenfeld, B. M., Raghuram, S. and Garud, R. (1999). 'Communication patterns as determinants of identification in a virtual organization'. *Organization Science*, 10(6): 777–790.

Wilkinson, R. and Hingston, P. (1991). 'Using the cosine measure in a neural network for document retrieval'. In *Proceedings of the 14th Annual International Conference ACM SIGIR Conference on Research and Development in Information Retrieval*, pp. 202–210.

Winograd, T. and Flores, F. (1986). *Understanding Computers and Cognition*. Norwood, NJ: Ablex.

Wong, S. K. M., Ziarko, W. and Wong, P. C. N. (1985). 'Generalized vector space model in information retrieval'. In *Proceedings of the 8th Annual International Conference ACM SIGIR Conference on Research and Development in Information Retrieval*: 18–25.

Wolpert, D. H. and MacReady, W. G. (1997). 'No free lunch theorems for optimization'. *IEEE Transactions on Evolutionary Computation*, 1(1): 67–82.

Yang, J. (2003). 'Web service componentization'. *Communications of the ACM*, 46(10): 35–40.

Yang, Q. Z. and Zhang, Y. (2006). 'Semantic interoperability in building design: methods and tools'. *Computer-Aided Design*, 38: 1099–1112.

Zack, M. H. (1999). 'Developing a knowledge strategy'. *California Management Review*, 41(3): 125–145.

Zack, M. H. and McKenny, J. L. (2000). 'Social context and interaction in ongoing computer-supported management groups'. In Smith, D. E. (ed.), *Knowledge, Groupware and the Internet*. Boston, MA: Butterworth-Heinemann.

Index

4Projects 55
7C model 49
80–20 software 61

access to information perspective 43
Aconex 55
active mediators 150
activity 56
adjacency graph 138, 139
AEC Building Systems Model 74
Alavi and Leidner: Perspectives on
  Knowledge Management 41, 43
algebraic model 67
alliance 13
  branding and marketing 169
  business opportunities management
    in 171
  definition 167–8
  innovation promotion in 173–5
  operations management 168–9
  sustainability of 171–3
ant colony algorithms 106
application service providers (ASPs) 36,
  55, 169
application-driven data integration
  147–8
application-driven semantic
  integration 148–9
Asite 55
AskMe 61
ATLAS 38, 74, 75, 81, 148
auto finding experts 60
AutoDesk 37, 55, 61, 72
awareness raising 180–1

bcXML 71
Bee algorithm 106
behavioural school 42
belief network 67, 69, 70

Bentley Systems 37, 55, 72
best of breed solutions 12
best practices 31, 32
BGRID 125
  component sizing 133–4
  controlling the search 134–5
  evaluation 135–6
  fitness function 128–30
  representation 125–6
  reproduction, crossover,
    mutation 126–7
  typical user session 130–2
BIM solutions 38
binary encoding 109–10
binary string 108
BIW Technologies 55
Boolean model 67, 68
Bricsnet 61
brief development 124
British Standards Institution (BSI) 37
broker/customer management 169
BS 6100 37, 71, 89, 91, 92, 148
building information modelling
  (BIM) 35, 74–5, 124, 148
building ontology 159
Building Research Establishment 26
Building SMART 148
bureaucratic cultures 180
Business Collaborator 55
business opportunity management 169
  in an alliance 171
business process reengineering (BPR) 32
Buzzsaw (AutoDesk) 54, 55, 61

CAD 36, 37, 38, 72
  4D 149
CadWeb 55
canonical genetic algorithm 107–8, 122
capability perspective 33, 43

case-based reasoning (CBR) 58–9
Causeway 55
change
    acceptance of 177–9
    types of 177–8
change management 36, 189
    in value creation 194–5
Chimaera 84
client briefing 124
cohesion 163–4
Collabor8online 55
collaborative search 60
COMBINE 14 148
    Integrated Data Model 38, 74, 75
combined research model 50–1
commercial school 41–2
COMMIT 74
commodification of knowledge 4
communities of interest 16
Communities of Practice (CoP) 16, 31,
    32, 39, 75, 188
community-based model 49–50
complex problems, examples of 102–3
    built environment design 102–3
    power generation 102
    strategic decision making for fire and
        rescue 103
composition/aggregation relationship
    93
computer support for co-operative work
    (CSCW) 56
concept level integration 90, 91
concept relationships 93–6
conceptual design 124
conceptualisation, definition 75
condition perspective of knowledge 33
CONDOR´ 74
confidence 182–3
confidentiality 19, 178
construction alliances 164–6
construction industry
    barriers to knowledge sharing and
        technology adoption 24–6
    data, information and knowledge
        needs 26–7
    essential attributes 29–30
    knowledge management generation
        in 32–6
    limitations of knowledge
        management 27
    research and development 28–9
    small and medium-sized
        enterprises 23–4
    structure 21–3

Constructware 55
continuous learning 164
corporate business processes,
    understanding 183
customer relationship management 169

data, definition 3–4
decision support systems 58–9, 105
demand-side knowledge management 32
Department of Trade and Industry 58
Diffusion of Innovations, stage model of
    (Rogers) 79–81
discipline 165
discussion forum 60
DIVERCITY project 149
document management systems 36–7
    traditional 53–4
document review 60
document type definition (DTD) 65
domain knowledge 26
Drama Theory 46
DXF (Drawing/Data Exchange
    Format) 37, 72, 148

Earl's schools of knowledge
    management 41–2
eCognos 61–3, 79, 87–8,s 96–100, 148
economic school 41–21
EDI 150
EDICON 150
EDIFACT standards 150
electronic document management
    (EDM) 53–4
electronic document management
    systems (EDMS) 34, 36–7, 186
email 190
embedded knowledge 4
embodied knowledge 4
embrained knowledge 4
encoded knowledge 4
encoding 108–11
encultured knowledge 4
Engineering and Physical Sciences
    Research Council (EPSRC) 29
European Commission 28
evolutionary algorithms
    distinguishing features 106
    reason for 101
evolutionary operators 137
    crossover 137, 141
    mutation 137, 140–1
    selection 137
evolutionary programming 106
evolutionary strategies 106

experience, value of 103–4
explicit ('codified') knowledge 4–5, 18, 38, 51
Express 78
Express-X 79
extended enterprise 13
extranets, project 54–5

face-to-face interactions 16, 39, 42, 47, 48, 164, 180, 190
FCA-Merge 84
feedback and evaluation 185
financial risk 165
fitness assessment 111–16
fitness function 108, 111–16
full text search 59–60
functionalist paradigm 42–3
FUNSIEC project 79
fuzzy model 67

generalisation/specialisation relationship 93
generalised vector 67
generalised vector space model 68
genetic algorithms 106, 107–8
  choice of encoding 109–11
  convergence and results 121
  crossover 118–19
  fitness function 111–16
  inversion 121
  mutation 119–20
  problem encoding 108–9
  selection using 117
genetic algorithms for design 123–44
  BGRID fitness function 128–30
  BGRID representation 125–6
  BGRID reproduction, crossover, mutation 126–7
  BGRID: typical user session 130–2
  building design 123–5
  component sizing 133–4
  controlling the search 134
  creating initial population 133
  evaluation 135–6
  evolutionary operators 137, 140–1
  further developments 136–7
  OBGRID and orthogonal buildings 137–8
  orthogonal building example 142–3
  results 143
  search results 134–5
genetic programming 106
globalisation 12
GraphiSoft 37, 72

groupware systems 55–6

Hamming cliff problem 110–11
harmony search 106
health and safety 187
heuristics 103
HTML 65
human networks in value creation 189–90

IBM Lotus QuickPlace 61
ICT
  literacy 184
  role in alliances 173, 174
index term selection 66–7
individual knowledge 26–7
Industrial Revolution 1
Industry Foundation Classes (IFCs) 35, 38, 74, 76, 78–9, 87, 148, 151
  model 71, 72
  Model Servers 38
  reasons behind low adoption of 79–81
inference network 67, 69–70
information, definition 3–4
information management environments 57–8
information retrieval 32, 63–4
information system (IS) 5, 52
information technology view of knowledge management 5
Initial Graphics Exchange Specification (IGES) 148
innovation in knowledge organisations 17–18
insight generation 51
intangibles 15–17, 188, 193
integration approach 146
integration operation 89
intellectual capital 15–16, 42, 189
  in value creation 193–4
intellectual property rights 178
intelligent buildings 145
Intelligent Design Assistants (IDeAs) 148
International Alliance for Interoperability 38, 74, 148
International Standard for the Exchange of Product Model Data 38, 74, 148
interpretive paradigm 43
inter-programme cooperation 28
interproject coordination 28
intranets, project 54
ISO 10303 38, 74, 148

ISO 12006 2 71
ISTforCE project 148, 149

Just in Time (JIT) methods 166

KACTUS project 83
KAON2 84
knowledge, ICT evolution and 52–3
knowledge-base generation and
    management 60
knowledge-based systems (KBS) 58
knowledge category model 42
knowledge conceptualisation and
    nurturing (second generation) 37–9,
    187
knowledge conversion 5, 32
knowledge creation 31, 48–51
knowledge, definition 3–4
knowledge environment 30
knowledge in action 45
knowledge industries 2
knowledge-intensive living systems,
    buildings as 157–60
knowledge management
    approaches 43–8
    reason for 1–3
    practical implications 5–6
knowledge management generation
    accounts 31–2
    criteria 34
    in construction 32–6
    knowledge conceptualisation and
        nurturing (second generation) 37–9,
        187
    knowledge perspective 34
    knowledge sharing (first
        generation) 32, 36–7, 163
    knowledge value creation (third
        generation) 39–40
    lifecycle focus 34
    socio-technical dimension 34
    underpinning ICT 34
knowledge management systems
    (KMS) 30, 52–70
    common and emerging
        functionality 59–61
    definition 53
knowledge mapping 47–8
knowledge organisation 2
knowledge perspectives 41–8
knowledge sharing (first generation) 32,
    36–7, 163
knowledge value creation (third
    generation) 39–40

latent semantic indexing 67, 68–9
learning 184–5
learning by doing 49
learning organisations 181
lessons learned 31, 32
lexical analysis 66
liability insurance 165
lifelong learning 181

map of ICT research 146–7
Matlab 107
McAdam and McCreedy: categories of
    model for knowledge management 41,
    42
merging operation 89
Meridian Project System 55
meta-data 36
Methontology 83
Metric Clusters method 94
Model Driven Architecture (MDA) 150
motivation 165

Network for Construction Collaboration
    Technology Providers (NCCTP) 55
neural network model 67, 68, 69
n-form organisation 18
non-overlapping lists model 67

OBGRID 136, 137–8
    adjacency graph 138, 139
    partitioning 137–8, 139–40
object perspective 33, 43
OIL 84
Oiled 84
OmniClass Construction Classification
    System (OCCS) 37, 71
OntoEdit 84
On-To-Knowledge methodology 83
Ontolingua 84
ontology 39
    architecture definition 87–8
    definition 75
    development methodology 82–3 85–6
    philosophical approaches 75–6
    product data vs 76–9
    requirements 81–2
    testing and validation of 96–9
ontology modules construction 89–96
    document cleaning 89–90
    integration index terms into core and
        sub-ontology 90–1
    keyword extraction 90
    ontology concept relationship
        building 93–6

Ontology Web Language for Web
  Services 153
OntoMorph 84
Open Semantic Infrastructure for
  the European Construction Sector
  (OSIECS) 79
organisation school 42
organisational change 18–19
organisational conversation 47
organisational culture 176–7
organisational goals 176
organisational knowledge 26
organisational learning 193–4
OSMOS project 74, 149
OWL (Web Ontology Language) 76–7,
  84
OWL-S 153

P3 platform 148
Palladio 33
participatory type of culture 180
particle swarm analysis 106
partitioning 137–8, 139–40
partnering 13, 17, 164–6
pattern discovery 50
personal knowledge 24
PGSL 106
Primavera 55
PrimeContract 55
probabilistic model 67, 69–70
problem-structuring methods 46–7
process perspective 33, 43
Process Specification Language 150
process-driven data integration 149–51
product cycles, shorter 12
product data 37
profiles or context-based search 60
project knowledge 26
project memory 56
Project, The 55
ProjectNet 54, 61
ProjectTalk 55
ProjectWise 55
Prompt 84
Protégé 84
proximal node model 67

RATAS model 38, 74, 148
RDF (resource description
  framework) 64, 84
real number encoding 109
real numbers 109
reflection-in-action 45–6
reflection-on-action 46

requisite variety, law of 44–5
research and development 28–9
resistance to change 3
reward systems 25
risk management 17
Robustness Analysis 46
rules of thumb 103

SABLE project 149
satisficing solution 125
schema theory 109–10
Schultze: perspectives on knowledge
  management 41, 42–3
search algorithms 106
search engines 105
search facility 59–60
search space 104, 1–5
SECI model 31, 32, 48–9, 50
security 19
semantic focus 146
semantic process-driven vision
  for the future of construction 151–3
  in practice 153–7
semantic relationship between
  concepts 93
semantic resources in construction
  sector 71–2
semantic resources selection 88–9
semantic search 60
sensemaking 43–4, 51
Sensus 83
service-based knowledge solutions for
  construction virtual enterprise 56–7
service-based software delivery 159
Seven S model 176, 194
shared conceptualisation 75–6
SilentOne 61
simulated annealing 107
skills gaps 183–4
Slimflor system 133–4
small and medium-sized enterprises 23–4,
  166–7
  alliances, knowledge needs for 168–9
  innovation in 173–5
smart alliances 159
smart components 159
smart home environments 145
smart materials 158, 159
social capital 16, 188–9
  in value creation 190–1
social constructionist model 42
socialisation view of knowledge
  management 5
Soft Systems Methodology (SSM) 46

solution formulation 51
spatial school 42
stakeholder engagement 179–80
Standard Fitness Method 130
state of mind perspective 33, 43
stemming 66
STEP 35, 38, 74, 76, 78–9, 99, 148, 151
stopwords elimination 66
strategic appraisal 51
strategic choice 46
Strategic Options Development and Analysis (SODA) 46
strategic school 42
supply chain 13
supply chain management 17
supply-side knowledge management 32
survival of the fittest 108
sustainability goals 187
sustainable alliances, criteria for 171–3
'sweep line' algorithm 137
SWOOP 84
Sword Group 55
Sword-CTSpace 55
syntactical integration 90

Tacit ActiveNet 61
tacit knowledge 4–5, 18, 24, 38–9, 51
tangible assets 15–17
taxonomy management 60
team
  definition 13
  identification 163
  virtual 13
team-building 190
technocratic school 41
Technology Adoption Model (TAM) 79, 80–1
technology in value creation 191–3
Technology Strategy Board 29
term categorisation 67
term frequency-inverse document frequency 91–2
text mining 63–70
  document type taxonomy 64–5
  documents and logical representation 64
  models for document semantics characterisation 67–70
  text operations 66–7
third party service providers (TPSP) 154–5

time to market 12
Tove 83
training 181–2, 192–4
transformation operation 89
translation operation 89
transparency 167
trust 16, 17, 19, 163–4, 165, 167, 182–3, 190

UniClass 148
uniform resource identifier (URI) 64
uniform resource locator (URL) 64
user authentication 36
user-centred education and training 184–5
user profiling 60

value creation 186–96
  change management in 194–5
  human networks in 189–90
  intellectual capital in 193–4
  reason for 188–9
  social capital in 190–1
  technology in 191–3
value management 17
vector model 67, 68–9
virtual alliance 178
virtual corporation 13
virtual enterprise (VE) 13, 56, 162–4
virtual enterprise clients (VEC) 155
virtual enterprise service provider (VESP) 154
virtual organisation 13
virtual teams 13–14
  in construction 162–4
  structure of 14
Vitruvius 33

W3C 153
web-based project management systems (WPMS) 55
Web Ontology Language (OWL) 76–7, 84, 153
Web Service Modelling Ontology 153
Woobius 55
WordNet 83
workflow functionality 60
workflow process 56–7
World Wide Web 32
WSDL 142

XML 64, 65